Computer Graphics in
Medical Research and
Hospital Administration

Computer Graphics in Medical Research and Hospital Administration

Edited by
R. D. Parslow
and
R. Elliot Green

Department of Computer Science
Brunel University
Uxbridge
England

Plenum Press · London and New York · 1971

Plenum Publishing Company Ltd.,
Donington House,
30 Norfolk Street,
London, W.C.2

U.S. Edition published by
Plenum Publishing Corporation,
227 West 17th Street,
New York, New York 10011

SBN: 306-30518-6
Library of Congress Catalog Card Number: 75-137741

PRINTED IN GREAT BRITAIN BY
BELL AND BAIN LTD, GLASGOW

Preface

The graphics terminal makes it possible for people who are not computer specialists to communicate with computers on an interactive basis, without the delay or inconvenience of working constantly through an intermediary. It provides a language of shapes or symbols (full graphics) or words and numbers (alphanumerics) which is understood by both man and machine. The visual output and input facility has considerably widened the applications of computers within the medical world, bringing their enormous powers of data handling and simulation to bear on solving problems in administration, patient monitoring and clinical analysis and research.

The purpose of this book is to provide examples of the work being carried out now in the U.K. and U.S.A., showing the applications of all types of installations—from small to very complex—for both administrative and research uses. It gives a brief overview of benefits already derived and of future plans; of hardware utilisation and of software approach; of problems met and of problems solved.

The intention is to acquaint executives and researchers in all branches of the medical world with the rapid progress being made in computer graphics and to stimulate thought on which way the technique can be developed to the advantage of all.

The material is based upon papers which were presented at the Computer Graphics 70 International Symposium held at Brunel University, Uxbridge, Middlesex, in April 1970. We should like to express our special gratitude to the authors for providing up-to-date information on their current work and such admirable illustrations; and also to D. White, Head of Computer Policy Branch, Department of Health and Social Security, U.K., and to Professor M. L. V. Pitteway, Professor of Computer Science at Brunel University, two of the session organizers at the Symposium, for their invaluable help in assembling the material used in this book.

v

Readers interested in looking further into the techniques and applications of computer graphics outside the medical world are referred to companion volumes by the same editors:

"Computer Graphics—Techniques and Applications" published by Plenum Publishing Company Limited, 1969 *and*

"Advanced Computer Graphics—Economics, Techniques and Applications" published by Plenum Publishing Company Limited, 1970.

ROBERT D. PARSLOW
RICHARD ELLIOT GREEN

Brunel University,
Uxbridge, Middlesex,
England.

September 1970

Contributors

R. E. BENTLEY, *Institute of Cancer Research, Royal Marsden Hospital, Belmont, Sutton, Surrey*

C. BRAUER, *Computer Programmer, Technology and Systems, Salt Lake City.*

D. W. COPE, *Digital Equipment Co. Ltd., Arkwright Road, Reading, Berks.*

D. A. FRANKLIN, *Computer Unit for Medical Sciences, St. Bartholomew's Hospital, London, E.C.1.*

H. GREENFIELD, *Assoc. Res. Prof. Computer Science and Asst. Res. Prof. Surgery, University of Utah.*

LOU KATZ, *Graphics Facility for Interactive Displays, Department of Biological Sciences, Columbia University, Morningside Heights, New York, 10027.*

CYRUS LEVINTHAL, *Graphics Facility for Interactive Displays, Department of Biological Sciences, Columbia University, Morningside Heights, New York, 10027.*

J. MILAN, *Institute of Cancer Research, Royal Marsden Hospital, Belmont, Sutton, Surrey.*

J. PERKINS, *National Institute for Medical Research, The Ridgeway, London, N.W.7.*

E. PIPER, *National Institute for Medical Research, The Ridgeway, London, N.W.7.*

JOHN C. A. RAISON, *Department of Health and Social Security, London.*

K. REEMTSMA, *Chairman, Department of Surgery, University of Utah Medical Center.*

B. RIDSDALE, *Chief Programmer, King's College Hospital, Computer Unit, London.*

F. G. TATTAM, *National Institute for Medical Research, The Ridgeway, London, N.W.7.*

J. WHITE, *National Institute for Medical Research, The Ridgeway, London, N.W.7.*

C. C. WILTON-DAVIES, *Royal Naval Physiological Laboratory, Alverstoke, Gosport, Hants.*

Contents

Chapter 1 The Visual Display Unit for Data Collection and Retrieval. *B. Ridsdale* 1

Chapter 2 Graphic Presentation of Clinical Measurement and Monitoring. *J. C. A. Raison* . 9

Chapter 3 The Use of Graphic Input and Output Devices Attached to a Small Computer for Planning Radiotherapy Treatment. *R. E. Bentley and J. Milan* 19

Chapter 4 Small Computer Graphics in the Physiological Laboratory. *C. C. Wilton-Davies* . . . 27

Chapter 5 Preliminary Analysis of Blood Flow Characteristics in the Abdominal Aorta by Computer Interpretation. *H. Greenfield, C. Brauer and K. Reemtsma* 35

Chapter 6 Computer Graphics in Molecular Biology. *L. Katz and C. Levinthal* 56

Chapter 7 Interactive Computer-Generated Stereoscopic Displays for Biomedical Research:

    Part I Development of Techniques. *J. White and J. Perkins* 71

    Part II Applications. *D. A. Franklin* . . 73

    Part III Development of the System for Molecular Structures. *E. Piper, J. Perkins and F. G. Tattam* . . 83

Chapter 1

The Visual Display Unit for Data Collection and Retrieval

B. Ridsdale
Chief Programmer
King's College Hospital
Computer Unit
London

This chapter describes part of the first phase of the King's College Hospital Computer Project. A medical information system is described which provides facilities for the on-line input and retrieval of clinical notes, and communications for certain instructions and test results.

KING'S COLLEGE HOSPITAL

The King's College Hospital Group consists of five hospitals all within 2 miles of each other. There is a total of 1,800 beds and 35,000 in-patients are treated each year. The annual operating cost of the Group is about seven million pounds.

The Hospital Computer Project is the first of the Department of Health's experimental projects. One of the aims is to provide an on-line real-time medical information system. The computer users at King's are doctors, nurses and other staff engaged in their usual work. The information system must therefore be an aid and not a burden, and must be sufficiently understandable to avoid the need for intensive training in usage.

The initial phase of the experiment is based on two medical wards within King's College Hospital itself, and deals with certain communications between these wards and a service department.

Data Input

The patients' clinical notes comprise a highly complex and often bulky volume of data. Prior to the introduction of the computer system these were all hand-written by medical staff.

The possibility of using a trained terminal operator to enter clinical notes was considered, but the time lag between the completion of the notes on the ward and their availability to the computer users makes this impractical. It was therefore decided that the users should interface directly with the visual display units. Various alternatives were considered. It was considered unacceptable, both for educational and for practical reasons to expect medical and nursing staff to type in their notes on the terminal. The use of a coding system would cut down the quantity of typing involved, but in view of the large vocabulary of the doctor, this was quite impractical. The branching questionnaire technique was therefore chosen.

By means of the branching questionnaire technique, the user is helped to record his notes by a series of questions. At each stage a questionnaire is displayed on the visual display unit screen. The user selects one of the possible choices, and dependent on this the system then displays the corresponding next questionnaire. By using this question and answer system the user quickly progresses through a hierarchy of possibilities, and with only a few responses has specified in detail what he wishes to record. In the following example the user is being offered the opportunity of commanding the system to register a patient (1), enter clinical notes (implicit command in 2–8), or to retrieve patient information (9). He makes his choice by pressing the key corresponding to his choice number (e.g. 1 for PATIENT REGISTRATION), and then pressing the 'transmit' key.

    ✝ 1 PATIENT REGISTRATION
         2 HISTORY
         3 PHYSICAL EXAMINATION
         4 VITAL SIGNS
         5 INVESTIGATIONS
         6 DIAGNOSIS
         7 TREATMENT
         8 FOLLOW UP NOTES
       9 INTERROGATION

When the first choice has been made a new questionnaire will appear which will request further information. For example, if

TREATMENT were chosen (by typing a 7) the following questionnaire would be displayed:

SINGLE CHOICE

TREATMENT

1 NURSING ROUTINES
2 CLINICAL PROCEDURES
3 FLUID BALANCE
4 VITAL SIGNS
5 PHYSIOTHERAPY
6 OCCUPATIONAL THERAPY
7 DRUGS
8 DIETS

If NURSING ROUTINES were chosen, the next questionnaire would display a list of the basic nursing routines that can currently be ordered using the computer system.

SINGLE CHOICE

TREATMENT   NURSING ROUTINES

1 BATHING
2 MOBILITY
3 CARE OF PRESSURE AREAS
4 FEEDING
5 ORAL HYGIENE
6 LAVATORY
7 POSITIONING OF PATIENT IN BED
8 MEASURES FOR RELIEVING PRESSURE
9 WEIGHING
X END OF BASIC NURSING ROUTINES

As each choice is made, the text associated with the choice (e.g. FEEDING if 4 were chosen in the above questionnaire) is added to a sentence which represents the user's path through the questionnaire.

Thus the user can build up a complex sentence, piece by piece, by choosing at each step the next word or phrase to be added to it. The sentence so far could be:

TREATMENT

*NURSING ROUTINES *FEEDING

....

and would continue with detailed instructions for the nursing treatment of the patient.

Lists of choices are only applicable where there is a finite set of
words suitable to the specific context.

When the above conversation reached the following point:

TREATMENT
    *NURSING ROUTINES *FEEDING *NEEDS
HELP WITH FEEDING   *STERILE WATER ONLY

    :............

the next questionnaire to be displayed would be a plain form:

TREATMENT   STERILE WATER ONLY
    GIVE] [MILLILITRES
      AT] [HOURLY INTERVALS
       ] [TIMES A DAY

The user would then fill in all the relevant 'boxes' in the form between
the square brackets]............[.

When this was complete the system would display the total sentence
for the user to check:

TREATMENT
    *NURSING ROUTINES *FEEDING *NEEDS
HELP WITH FEEDING   *STERILE WATER ONLY
:GIVE 30 MILLILITRES—AT 1 HOURLY INTERVALS

When he had signified his approval the sentence would be filed.

Features of the Branching Questionnaire Technique

The following are some of the features of the branching questionnaire technique of 'writing' sentences, which might be used for comparison with other techniques.

The branching questionnaire system 'guides' the user by presenting
him at each stage with all the possible phrases which he may now add
to his sentence. It is not necessary to abbreviate the phrases merely
to minimise typing, as he indicates his choice merely by a number.

In 'writing' a long sentence the user is saved the frustration of
finding at the end that a mistake was made early on. Each choice or

entry into a form is checked when received by the computer and is returned for correction if found to be erroneous.

It is necessary to provide training in the use of the terminal, but using the branching questionnaire technique no memorisation of standard terms is required.

The syntax of the final 'sentence' is prescribed by the vocabulary on the questionnaire and the order in which they are presented. Thus the set of questionnaires must be able to cater for a very large number of foreseeable circumstances. There are currently almost 4,000 of them available on disc for use by the real-time program.

The speed at which a user can 'write' an input language sentence is likely to have a large bearing on the acceptability of the language in such areas as admissions clerking and medical recording. By allowing the user to specify whole phrases merely by indicating a choice number, work is minimised. Further, when a number of sentences need to be recorded on the same overall subject, the user may 'hold' a particular questionnaire from one sentence to the next, therefore avoiding the need to make the same 'high level' choices repeatedly. Thus in the above example, if the 'CARE OF PRESSURE AREAS' sentence was followed by a sentence regarding 'WEIGHING', it would not be necessary to go through the choices of 'TREATMENT' and 'NURSING ROUTINES' for a second time.

By presenting questionnaires containing only the acceptable words at each stage of the sentence, the branching questionnaire technique provides a built-in safeguard against the use of valid (and invalid) words in the wrong context. A double check is assured, as the resulting sentence is displayed for the user to approve before action is taken.

Other uses of Branching Questionnaires

Using the branching questionnaire technique is not only limited to the input of data; an initial set of questionnaires allow the user to select the on-line facility that he requires, and the nursing sister uses a set of questionnaires to record instructions for the treatment of patients. These instructions are subsequently used by the computer system to produce work lists for the nurses.

Retrieval

Once a complete sentence has been constructed using the branching questionnaire technique, and then approved by the user, it is of course filed. It immediately becomes available to an authorised user at any VDU which is connected to the real-time program. An individual patient's medical record is retrieved by identifying the patient and then viewing the notes sentence by sentence in chronological or reverse chronological order of input. An enhancement will be added in the near future to allow the user to select a specific part of the individual patient's record (e.g. the physical examination), and to scan only these records.

System Commands

There are occasions when the user may wish to change from the normal sequence of events. For example, when entering notes he may accidentally select the wrong choice or delete the contents of the screen, or he may find the display system inadequate for entering some special type of data. To cope with circumstances such as these, a set of system commands are available. In the above examples the user would transmit a 'B' (for Back) to request the system to go back to the previous questionnaire, transmit 'A' (for Again) to cause the same display to be repeated again, or transmit 'N' (for Narrative) to allow him to add free text to be appended to his sentence.

Content Independence

It may be noticed that the format of the different types of questionnaire available in the system is independent of the content. In fact the branching questionnaire technique may be used for recording notes, instructions or requests, and although different questionnaires are needed, they are all handled by the same part of the real-time program. Because the functioning of the program is completely independent of the content of the questionnaires, the questionnaires can be easily added to and amended. The design and development of displays has been the responsibility of the representatives of the users themselves. A team of doctors has designed all the (almost 4,000) questionnaires for medical notes, and a nursing team has designed the procedures and the questionnaires for nurses.

A suite of programs is available for handling the 'housekeeping' of groups of questionnaires. These programs allow the teams to store and print their own groups, to insert, delete and amend questionnaires, and to link and load the groups so that they may be tested on-line.

System Response Time

Some attempts have been made to estimate the longest acceptable time between the user's request and the visual display unit's response. Experience has shown that users become highly frustrated when the response time stretches far above one second, and as they become more familiar with the system the critical response time becomes shorter.

On the branching questionnaire system the effect of this can be somewhat lessened by allowing the user to remember frequently used paths through the display system, and to enter his sequence of choice numbers as a response to the first questionnaire. The user may be prepared to take slightly longer on certain operations if the result is better, but this still leaves little latitude for the system response time, as the user's own response time may be more difficult to improve.

Implementation

Before the current system was developed the branching questionnaire technique was simulated using a single visual display unit connected to a PDP-8/S computer. When this first became available in 1968 it was possible to do some of the development of questionnaires and to appraise the user reaction. Much useful experience was gained from this experiment and a considerably improved program was available on the 1905E in August 1969. This provided a base for large-scale development of branching questionnaires. The first phase of the pilot scheme became available in February 1970, having benefited from the experience of both earlier experiments, and the first phase of enhancements is already well in progress. This includes the development of an education system which will work in parallel with the on-line program on the words, a fast means of patient identification by ward list and an improved system of on-line retrieval which will allow a far greater selectivity of relevant records.

Conclusion

This chapter has described some of the facilities provided by the first phase of the King's College Hospital Medical Information System. These facilities have been provided after experience with previous trial systems, but it is not expected that development will end here. The experiences of program design, hardware utilisation,

questionnaire development, and, most important, user satisfaction are now being channelled into the production of the next phase of the project, which will provide services to further areas of the hospital.

Chapter 2

Graphic Presentation of Clinical Measurement and Monitoring

John C. A. Raison, M.A., M.D.
Department of Health and Social Security
London

In a unit for intensive treatment of critically ill patients, a computer can make very significant contribution to the measurement, monitoring, and data collection system. The demands made upon the methods of data presentation are probably the most complex at present required in the health services. It is the purpose of this chapter to describe this facet of data display, arising in one of the earliest practical applications, and thence to derive some conclusions of future clinical requirements.

THE INSTALLATION

An IBM 1801 (32K word, 2 microsec memory, a 2310 3-disc storage unit of half a million words each disc, and other peripherals) was installed as part of a clinical development collaboration between the Heart Research Institute of the Pacific Medical Center, San Francisco and IBM Inc.[1] [2] Up to 6 bed positions could be continuously monitored, with facilities for special measurements at the remaining 8 in the intensive therapy unit, as well as the cardiac catheterization and exercise laboratories. Of necessity the computer was housed in a separate building 1,000 feet away.

Four methods of graphic display were provided by the system. The first was a conventional analog split beam oscilloscope operated directly from physiological transducer amplifiers, before signal input to the computer. An important purpose of this was as part of the fail-soft systems arrangement should the computer go down.

9

It was discovered that a heavy reliance was placed by experienced staff on the continuance of previously well-known analog signals, for arterial pressure and electrocardiogram. Attempts to secure confidence in the calculated and displayed values obtained by computer, by diminishing the analog screen to very small size met with sturdy opposition and doubt that the computer might not be processing a true biological signal. Interestingly, there was no clamour for a similar analog display of signals which were required for the first time when the computer was introduced, to show airway flow and pressure, which were of no less importance: the processed data alone was accepted by nurses.

The second provision was a twin channel, heat stylus type, chart recorder for hard copy display of analog signals, either current or retrieved from the data log. The quality of signals reproduced from store was extremely good. Of considerable value in research studies, its use in clinical care was not as frequent as anticipated.

The third technique was the use of an incremental XY plotter (Cal-Comp plotter, California Computer Products, Inc., Anaheim, California) which produced charts of all measured variables and of derived values for each 24 hour period. These were used each morning for the main ward round by the full clinical team. And they were of very great help in research and development of improved monitoring techniques, in forming preliminary opinions of potentially significant trend changes, and of narrowing down to specific areas for study the immense quantities of digital data acquired by computer based monitoring.

The fourth display method provided most of the routine operational information. Display programs were stored on disc. Up to 3 displays were generated digitally and 'simultaneously' (by multiprogramming) by the computer and passed through a D/A convertor to 3 small Tektronic storage oscilloscopes. A closed circuit television camera focused on each of these, and further analog transmission to the bedside 11 inch monitors was by a 1,000 ft direct line. The visual quality of this signal was acceptable for alphanumeric and redrawn analog signals, but burning of the tube gradually diminished clarity.

When displays were virtually continuous throughout the day, storage 'scopes burnt to an unacceptable state for reading in 2–3 months. As a result, it was arranged that displays should be removed after 30 sec unless a hold button or switch demanded its persistance. This increased tube life by a factor of 2–3.

At the start of clinical operation, alphanumeric tabulations of data were produced. The various medical measurements were set out in horizontal lines, in columns according to the times of analyses, and with the most recent values at the *left* of the screen, close to the relevant descriptive term for each measurement. Up to 12 lines of data, and up to 6 columns could be displayed and there were various programmed techniques for push button command to retrieve different standard display formats, or to divide the screen so as to display information concerning two patients simultaneously.

It was observed that, although nursing and medical staff made regular use of the data so presented, traditional 'vital signs charts' (e.g. of blood pressure, pulse and respiration) were still prepared, using the values display on the screen. Clinical review of progress now seemed to include some mental correlation of displayed digital data with the prepared charts. As a result, two further stages of display development took place. The first was an *x–y* plot displaying any one of three variables over any desired time base from 1–24 hours. One of the standard type characters (dot, asterisk or dash) was used for each measurement shown. No vector connection was made between isolated values, and the most recent appeared at the *right* of the screen. Whenever a display was to be made, after routine analysis events, or when otherwise requested, the coordinate axes, with maximal and minimal values, were first laid on the screen. The display program then searched the data, logged on disc, for the required values. As each group of values for a given time was retrieved, it was displayed, so that the presentation had a dynamic character which enabled the observer to overcome most of the difficulty in discriminating and linking an irregular series of independent values. The subsequent development was to limit the left half of the screen to this form of graphic display, and to provide on the right side a column of the most recent numerical values of variables, with their descriptors in words besides them. Thereafter these became the standard means of clinical review of progress. The ability to obtain a plot of short or long time base by single command appeared to be of great value. This pattern of display proved to be the most acceptable and useful for treatment purposes.

It was inevitable during such early trials that much of the original or calculated data was displayed in stylised manner. However, whenever a clearly or probably established trend occurred, programming provided for a *verbal* display highlighting the change. Early examples were 'decreased pulmonary compliance' and 'arterial catheter damped'.

Displays of higher priority were made to flash. The impact of this seemed rapidly to diminish on use. For highest priority messages, flashing lights and alarm sounds were provided. Effectiveness in obtaining staff response was proportional directly to intensity, and inversely to frequency of probable false alarm content.

At a later stage, a fuller graphic display was required. One of the physiological signals measured the change of pressure in the patient's airway, plotted against change in volume of air in the lungs, forms a loop during inspiration and expiration. In the particular circumstances of care, only the inspiratory portion is of importance for ventilatory treatment. Presentation of only the rising volume or pressure phase was achieved by programming. The mode of display was on an XY plot, using a constant vertical axis for volume, but moving the zero point of the horizontal (pressure) axis an indicated amount to the right for each successive inspiration. Each intaken breath was plotted immediately on completion, could be compared with its predecessor to identify the degree of change obtained as adjustments were made to the ventilatory machine. This type of 'edited' analog display, which could also be stored on demand, was a recognised adjunct to care and observation.

As originally devised, the system acquired most of its information directly in analog form: an 1816 keyboard-printer was provided for input of alphanumeric data, as well as some data output. For reasons of noise, inconvenience of position (a single station centrally placed between two rooms) and inhibitions about typing, this method did not gain favour for input. This fact contributed greatly to the decision to provide new display units for each patient position, each made up of a small 16 button keyboard and 11 inch standard television receiver. As a result, display techniques became a part of *input* procedure as well as data output. A branching system of presentation was used; each screen provided up to 10 lines of numbered choices, each of the latter leading to another 10 choices: for most purposes only three nodes were necessary, although the extensive nursing attention notes demanded further branching. Each display was sub-numbered from its parent, providing the experienced operator with even faster access to the required display by use of the appropriate series of display numbers in immediate succession (e.g. pushing "1, 3, 3, end, transmit", rather than "1, end, transmit, display, 3, end, transmit, display, 3, end, transmit, etc.").There was no provision for overwriting on the storage 'scopes used, so that data added called for regeneration of the updated display for confirmation before formal storage. The use of this mode, and the provision of relatively mobile display units which could be placed in the most suitable place for needs of a particular patient's nursing condition, rather than a single display on a console, increased the manual input of important data by a very large amount.

The introduction of a computer based measurement and monitoring system to the acute nursing environment brought a wide range of developments bearing on many aspects of the work. It was necessary to effect changes and subsequent developments in small elements, each preferably operating in parallel with the old method, until evidently reliable to the staff concerned. It was also clear that considerable modification of displays and formats would be required as working experience with the system developed. Still later, as new methods of observation or analysis were added, further modifications of display were needed. There were three levels of control in securing change of display. The first was the simple console button or switch commands given to secure any one display. This leaned most heavily upon shrewd judgement of the most appropriate apposition of information, and the most commonly requested information to secure minimal computer activity with maximum clinical usage. As mentioned earlier, some graphs of data were appropriately shown with short time bases (usually 4 hours) at some times, and at others with the values over 24 hours shown. It was convenient to make this choice a switch position command. The second level of display control was by keyboard instruction: considerable variation of format, or even of the specific items of data retrieved for display, could be obtained in response to coded signals. Such a facility tended to be used only by the 'specialists' in the unit, such as the head nurse, unit director or research worker. The third means of display alteration was by substitution of 1 or more cards in the original card program for each display which had been preserved. Such changes were effected in the course of display development, by the team in consultation. Great importance in terms of maintaining user confidence, was attached to being able to effect all such changes very rapidly after agreement—usually within 24 hours, but with the facility for nearly instant reversion if found unacceptable on clinical introduction.

DATA DISPLAY REQUIREMENTS

This experience furnishes evidence from which it is possible to outline some optimum requirements of a display system when used for intensive therapy or in other similar clinical situations such as operating theatre or special investigatory laboratories.

(1) Display methods must have the best reading characteristics available at the time for what can be afforded. Thus, oscilloscope presentations must be flicker-free, clear in daylight and artificial lighting, and of good legibility. Any hard copy produced must be similarly clear, smudge proof and clean to handle, and any fixing method required should retain the picture for as much as a year.

(2) The system must be capable of
    (a) Repetitive presentations. Since standard formats relating to routine analyses will be the most common appearance, any loss of information or clarity due to tube burning of areas most illuminated must be prevented or minimised.
    (b) A system of urgent presentations (alarms).
    (c) Hard copy graphics.

(3) The bulk of the display equipment should be small in relation to the size of screen. Apparatus congestion in the clinical situation is already a serious problem; in particular the anterior–posterior dimension should be diminished. The unit should be capable of separation from any other of the computer units, consoles, or any other computer and of mounting in several different positions according clinical needs of particular nursing situations. Since it will sometimes be desirable to read the screen at a distance, characters of 15–20 mm height are indicated. Similarly, the keyboard should preferably be detachable from the display so that if a nurse is obliged to attend her patient from a particular position for a period of time, it may be moved there easily and quickly. It should comprise no more than 16 keys, easily depressed and under the span of a single hand.

(4) Display characteristics should comprise alphanumeric characters and simple graphics. Experiment has shown that the loss of resolution suffered by the summation of storage oscilloscope character and TV raster lines is acceptable. Detailed graphic resolution, e.g. for projection of radiographs, although perhaps desirable in an ultimate system, is of relatively low priority. We have no evidence nor reason to think that colour display would greatly improve function.

(5)   There must be a data input and editing facility for interaction with the display. Most data input is of digital information to standard formats. The provision of a display cursor and movement buttons with a keyboard may be sufficient for entry and editing. Preferable would be the programmed movement of the cursor from one entry (or correction) space to the next: usually input displays will group associated pieces of information, all of which will be required for entry on each occasion. Elaborate algorithms for jumping with the cursor are probably inessential, therefore; rather, simple progression through the series of data values should not be irritatingly time consuming, even for editing. More costly facilities such as light pen and touch wire require experimental evaluation before any advantage can be claimed.

Although alphanumeric tables and graphic displays will be the stock-in-trade of any acute care computer system, it should be one of the important functions of such a system that a large measure of correlation and logical deduction is made by it, and that this data reduction should more and more convey output information as interpretations in *word form* of condition and required action. Such information has varying degrees of importance, and a hierarchy of indicants of this should be established. To reserve a portion of the screen for such notices is undesirable, because of concomitant crowding of routine data. It is preferable that when such events occur, their presentation should result in a new display of diagnosis and supporting data. There may be a place for some of the information being made to flash. It is suggested that a not disagreeable sound, a 'dulcet tone', with perhaps illumination of a coloured bulb should accompany a priority signal. Above this, critical alarm information should be conveyed by more demanding light and sound.

(6)   Analog signals must be displayed, some frequently, others on demand. TV raster can be tolerated. It is possible that the call for these will diminish with increasing reliance on a system, and the development of the practice of 'exception reporting' rather than unselected data presentation.

(7)   Flexibility to vary display format. Apart from the relatively inexpensive means this provides for variations to suit the wishes of individual medical teams, there is a valuable research tool here. One approach to making computer based monitoring more significant is to explore the correlation between different measurements and clinical phenomena. A simple form of such multivariate analysis is

to project any desired group of variables as measurements along axes. Three dimensional projection is not required. It is possible to link graphically the coordinates, one for each variable, of a number of axes radially directed from a common centre. Since it is also possible to pulse the group of variables from each separate timed analysis, one may produce an apparently stable or changing spider's web pattern. Man's pattern recognition here may suggest correlations more quickly than by more precise methods of analysis.

(8) Signal switching and slaving of displays. Displays are required close to the patient under observation, but at a central nursing station it should be possible to obtain any display of any patient under observation. Similar facilities should exist at a number of other places, such as a doctor's office or night station, or the operating theatre. This includes transfer of analog signals. Voice communication between these is desirable for conference about displays.

(9) Closed circuit television observation of patients, which may be considered a form of data display, has been proposed: a few coronary care units use it, but its value is still much in question. The advantage of colour television, even ignoring cost, is even more suspect.

It is possible to conceive a single graphic display for all purposes in intensive therapy. The wiring arrangements would be complex. Analog signals must be transmitted directly, from amplifiers, for reasons of fail-safe: other displays would stream from the processor. Switching ability should exist at the bedside, a central station, and under computer control. From the user's viewpoint, there is no outstanding preference for either storage oscilloscope terminal or the more familiar computer visual terminal: developments in other forms of graphics may be as acceptable. These may include a means of generating hard copy by fixing the originally transient display. The analogy is provision of a polaroid camera attachment to computer displays. One advantage of such techniques over the existing methods of hard copy productions, is the ability to show digital and verbal information on the same graphic plot. At present only the most expensive, noisy and relatively slow plotters have this facility. Such systems, however, would probably not eliminate the need for a single separate piece of equipment capable of more complex displays for a limited number of purposes. For the present, however, the view is advanced that a graphic system for intensive therapy areas should be planned on the modular basis, with separate displays for analog signals and computer originated material separated keyboards, with adequate switching and slaving facilities.

ACKNOWLEDGEMENTS

Most of the clinical experience related herein was developed with the collaboration of colleagues of the Pacific Medical Center, San Francisco, Drs. J. J. Osborn, J. O. Beaumont and F. Gerbode, and others, aided by USPHS grant HE:06311: it is very warmly recognised.

REFERENCES

1. Osborn, J. J., Beaumont, J. O. Raison J. C. A. Russell J. A. G. and Gerbode F. (1968). Measurement and monitoring of acutely ill patients by digital computer. *Surgery* **64,** 1057.
2. Osborn J. J. Beaumont J. O. Raison J. C. A. and Abbott R. P. (1969a). Computation for quantitative on-line measurements in an intensive care ward: in *Computers in Biomedical Research* vol. 3, ed. Stacy, R. W. and Waxman, B. Academic Press, New York (pp. 207–237).

BIBLIOGRAPHY

Archer, L. Bruce	The Structure of the Design Process. USA Department of Commerce Report No. PB 179321.
Bolt, Beranek & Newman Inc.	Development of Notation and Data Structure for Computer Applications to Building Problems. Report 1398. Bolt, Beranek & Newman, March 1967.
Britch, A. L.	Use of Computer Stored List Connected Models as a Design Tool. International Council for Building Research, Studies and Documentation Report 13A, Oslo 1968.
Cropper, A. G. and Evans, S. J. W.	Ergonomics and Computer Display Design. The Computer Bulletin, Vol. 12, No. 3, July 1968.
Hawkes, Dean Stibbs, R.	The Environmental Evaluation of Buildings. Land Use and Built Form Studies Working Paper No. 15.

Jones, C. B.
A Program to Calculate and Display Daylight Factor Contours for Unglazed Windows.
Computer Technology Group Report 67/1.

Newman, W. M.
An Investigation of Methods using Computers for Processing and Storing Architectural Design.
Unpublished Ph.D. thesis, University of London, 1968.

Newman, W. M.
An Experimental Program for Architectural Design.
Computer Journal, Vol. 9, No. 1, May 1966.

Purcell, P. A.
Job Architect Analysis: A Study of the Job Architect's Role, for a Computer Aided Design Project.
Paper given to International Symposium on Man–Machine Systems.
Cambridge 1969. IEEE Conference Record No. 69C58MMS.

Singleton, W. T.
Current Trends Towards Systems Design.
Ergonomics for Industry, No. 12. Ministry of Technology 1966.

Sexton, J. H.
DISGOL—A Macro System with a Data Structure Aid for Graphical Display Programming.
Internal Paper: National Physical Laboratory, London 1969.

Chapter 3

The Use of Graphic Input and Output Devices Attached to a Small Computer for Planning Radiotherapy Treatment

R. E. Bentley and J. Milan

Institute of Cancer Research, Royal Marsden Hospital
Belmont, Sutton, Surrey
England

and

D. W. Cope

Digital Equipment Co. Ltd.
Arkwright Road, Reading, Berks
England

INTRODUCTION

The planning of treatment by radiation therapy is essentially a graphical problem. The object is to arrange X-ray fields in such a manner that regions of diseased tissue receive a high uniform dose while minimising the dose to certain critical organs such as the spinal cord. The practical difficulties of developing an acceptable computerised method are

(i) minimising the amount of work involved to input the data, and

(ii) obtaining results in a form which are readily accessible and easy to interpret.

In most radiotherapy centres, treatment by external X-ray beams is planned by a manual calculation. A treatment with two X-ray beams involves taking a chart of contours of equal dose drawn on tracing paper (known as an isodose chart) and overlaying it on a similar isodose chart for the second beam. Lines of equal dose for the two fields are then drawn in by hand by noting where various contour lines intersect. For each additional beam used in the treatment, the whole tedious process must be repeated.

Numerous computer programs, operating in batch mode, have been produced over the last few years to reduce labour and improve accuracy. In general, these programs require many laborious measurements followed by hand punching of paper tape or punch cards before the computation can begin. Furthermore, in many cases, a graphical output is obtained only on a line printer, by improvisation in a manner for which the instrument was not designed.

This chapter reports the programming of a PDP 8/1 for radiotherapy treatment planning. This allows the user to sit at the console of the machine and, by using graphic devices, obtain an immediate response. The intention is that it may be operated by staff with no special knowledge of computers. The basic concepts are somewhat similar to those applied to the Programmed Console designed and developed at Washington University, St. Louis, but significant improvements in hardware and software have been made.

One of the most important advantages of the digital computer in this field is its ability to readily repeat plans with differing X-ray beam configurations so that the best may be selected, generally choosing a new arrangement of beams from the results of a previous trial. Programs written to operate off-line, without graphic devices, do not offer this facility. In our system, data is input directly from a rho-theta position transducer and results are output to either an oscilloscope screen or a digital $X-Y$ plotter. The former gives a rapid response and the latter provides a permanent record.

MACHINE CONFIGURATION

The hardware required is one PDP 8/1 with 8K core (12 bit words), dual magnetic tape unit ('DECtape'), analog input channels, rho-theta co-ordinate transducer (Bolt, Baranek and Newman), Tektronix 611 storage oscilloscope and a 320 mm wide digital $X-Y$ plotter (Computer Instrumentation Ltd.). A hardware multiply-divide unit is virtually essential to reduce the operating time to an acceptable level. An overall view of the system is shown in Fig. 1.

OPERATION OF THE PROGRAM

This may be divided into four principal parts:

(1) Input of anatomical data for each patient.

(2) Generation and storage of X-ray beam data.

(3) Combination of dose distributions for beams placed in any position and orientation, followed by a search for the requested isodose levels.

(4) Output of the patient outline, positions of the beams and the resulting dose contours on an X–Y plotter.

Fig. 3.1.   Overall view of the system.

Stage 1—Anatomical data

The rho–theta position transducer is used to trace around a previously drawn cross-section of a patient (Fig. 2). The transducer produces voltages proportional to angle and radius which are input to the computer through an analog to digital converter. The program then converts the data to X–Y co-ordinates which are stored at 3 mm intervals for the whole outline. In addition to the outline, the position of diseased tissue, particular organs and other items of clinical interest

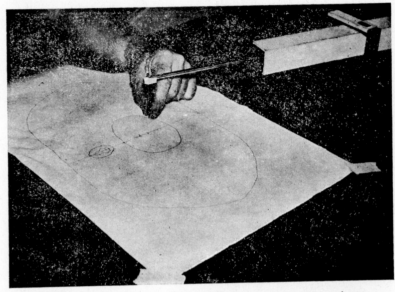

Fig. 3.2. Rho-theta position transducer used to trace a patient cross-section.

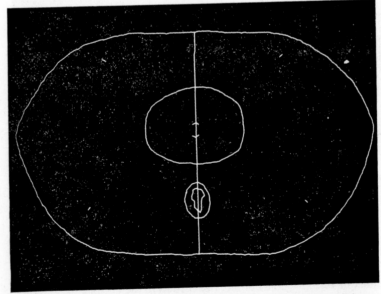

Fig. 3.3. Copy on the oscilloscope of the object traced with the position transducer.

may be similarly traced, digitised and stored, subject to a limit of about 750 individual points. During the tracing operation, a copy of the object being traced appears on the screen of the storage oscilloscope (Fig. 3). The digital data is transferred to DECtape for perpermanent storage and recall in stage 3 below. About 150 different outlines may be stored on one reel of tape.

Stage 2—X-ray beam data

There are many methods of obtaining the basic data for a particular beam, fully described in numerous papers on radiological physics. It would be out of place to give a detailed discussion here. The loading of appropriate data on to magnetic tape has to be done once for each beam. This is initially time-consuming but once a library of beams has been created, it may be used for many plans. Data for about 300 beams may be stored on each reel of tape.

Stage 3—Basic planning program

This is the most complex of the four stages. The original version of the program used 4K of core for the program and 4K for the matrix of doses to be calculated. Thus, each of the principal subsections (i) to (iii) below required overlays in core. Consequently, there was a delay of several seconds through tape action each time the user proceeded from one section to the next. More recently, we have shown that core overlays can be avoided by reducing the size of the output matrix to 1K without significant loss of accuracy on the screen. However, 4K is always used for output to the plotter and this does involve some tape action. The separate subsections of this program will now be described more fully:

(i) The outline and internal structure obtained during stage 1 is called the DECtape and displayed on the oscilloscope screen in storage mode. Although the individual co-ordinate data is stored at only 3 mm intervals, a continuous outline appears on the screen. This is done by using the so-called 'short vector' facility in the control system. X-ray beams are called from the second DECtape and appear as T-shaped objects (Fig. 4) which may be moved across the outline using 3 knobs (for X, Y and θ) connected to the analog channels of the machine. The 'write through' capability of the Tektronix oscilloscope

c

Fig. 3.4. Oscilloscope display of the patient outline and internal structures prior to matrix calculation. The "T" bars represent the position of each beam.

Fig. 3.5. Oscilloscope display of isodose contours produced from the beam arrangement of Fig. 4 after matrix calculation.

is used to allow the beams to be moved with a stationary picture in the background. The beam Ts require refreshing as in a conventional oscilloscope display. The same three knobs are used for all the beams and control is switched from one beam to another by typing on the teletype keyboard the number of the beam to be controlled.

(ii) The dose matrix for each of the required beams is called from DECtape. This is then translated from the beam co-ordinate system to the patient co-ordinate system using the knob settings from (i) above. Matrices for all the beams in the plan are summed. Using the extended arithmetic unit and the new program based on a 1K matrix, the calculation takes about 2 sec per beam.

(iii) Results are normally presented in the form of contours (iso-dose curves). If a computer method is to be successful, it must produce results in a familiar form. When the above stage (ii) is complete, the user may request any contour to be drawn on the screen by typing its value on the keyboard. The program then commences a search of the matrix and the appropriate contour line appears on the screen, using the storage mode of the oscilloscope. It takes approximately 1 sec to produce each line. Figure 5 shows a typical 4 field plan with the 10, 40, 80, 100, 120, 140 and 160% isodose levels.

If the dose distribution is considered to be unsatisfactory for clinical use, it is simple to revert to stage 1, change one or more parameters and repeat the calculation. The variables include beam position, angle and intensity and also certain constants appertaining to the beam itself.

Stage 4—Output to a Digital X–Y Plotter

The oscilloscope is ideal when the system is used in a trial mode but it is not very satisfactory for generating a permanent record. The radiotherapist requires a full-scale copy of a plan before commencing a course of treatment. When an acceptable result has been obtained on the screen, the plotting program may be started. This is a similar procedure to that used in the oscilloscope program but a different algorithm is used to maximise the efficiency of the slow mechanical device. The plotting time varies according to the amount of data which has to be drawn. A typical case shown in Fig. 6 takes about 4 minutes.

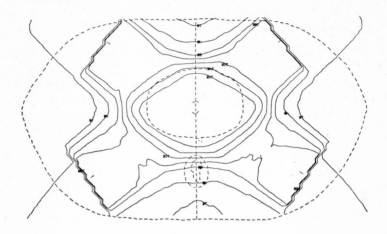

Fig. 3.6.   Digital X, Y plotter output of the isodose contours of Fig. 5.

SUMMARY

This chapter describes a medical application of a digital computer which becomes viable when graphic input and output devices are connected on-line to the machine. It simulates as closely as possible the familiar manual technique which has been in use for several years for summing the dose distribution of X-ray beams. It has greater speed and accuracy, saves tedious labour and allows a more flexible approach to the problem.

Chapter 4

Small Computer Graphics in the Physiological Laboratory

C. C. Wilton-Davies, M.A., M.B.C.S.
Royal Naval Physiological Laboratory
Alverstoke, Gosport
Hants.

When the benefits of using a computer in physiological research first became apparent at RNPL, we were faced with a choice between a data link to a large, central computer or a small computer on site. As our problems were mainly of the type involving a lot of data and a little computation, we chose the small computer. The specific machine was a LINC–8, manufactured by DEC.

As a government establishment, we felt ourselves to be under a certain obligation to buy British, but other factors eventually ruled out this possibility. The price of the hardware was one factor, but the overriding factor in our case was the quantity and quality of the software. Many other physiological laboratories in various parts of the world were using LINC–8's, so that there were already a number of programs in existence for doing some of the jobs that we wanted to do. As well as this, the whole tone of the software was oriented towards conversational access and user/machine interaction.

Judging by the experience of some laboratories using roughly equivalent British hardware, we have saved three to four man–years of programming time, and the machine had produced some useful results within a week of installation. Naturally, the supplied software package did not fill all our needs, and we have enlarged the program library both by our own efforts and from the very active users society. One hopes that the casual attitude of British manufacturers towards software is becoming a thing of the past, but there was little evidence of this in the biomedical sphere two years ago.

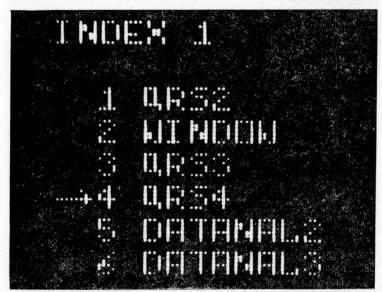

Fig. 4.1. CRT display of a part of the program index. The operator has selected the program "QRS4" for loading.

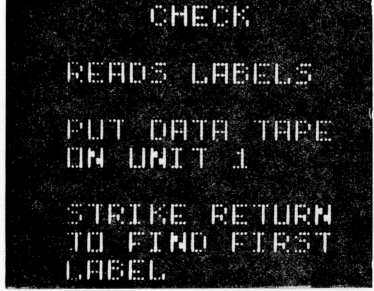

Fig. 4.2. CRT display during a program, asking for a specific operator action.

THE INSTALLATION

For the benefit of those unfamiliar with the machine, the LINC–8 has a core store of 4,096 12-bit words, two small magnetic tape units, a CRT display with alphanumeric and point-plotting facilities, an A-to-D converter and a number of relays for control of experiments. We have added an incremental plotter as an extra.

The CRT display enormously simplifies access to the computer at all stages of a job. A display of the contents of the index on a program tape (Fig. 1) is a quick way of locating a wanted program or rejecting the wrong tape, and eliminates many of the problems associated with keeping and accessing printed records. A display of text during the editing of programs provides much faster access than a printout from a teleprinter, is much quieter than a line-printer, and again saves the production of much unwanted paper.

During the running of a program, the CRT display not only provides conversational access to the machine (Fig. 2), but serves as a check on the data as it comes in and at intermediate stages of computation. This is particularly useful in the earlier stages of program development, and during very long experiments when the CRT display can give to those staff without computer training a reliable indication of whether the machine is performing satisfactorily.

GRAPH PLOTTING FACULTY

It has been mentioned that our problems generally involve rather a lot of data. The condensation of a great mass of data into a form which is readily comprehended is made possible largely by the use of the incremental plotter. The obvious way of doing this is the production of graphs, since the presence or absence of trends and variations is more readily apparent when data is in a graphical form than when it is presented as a printed table. Since we are too small an establishment to justify the full-time employment of a draughtsman, the value of this graph-plotting facility is considerable even in those cases when the amount of data is small (Fig. 3.).

As an example of what has been described, the computer analysis of the electrocardiogram is typical. Every time the heart beats, the action of the heart muscle is accompanied by the production of an electrical potential which may be detected through electrodes placed

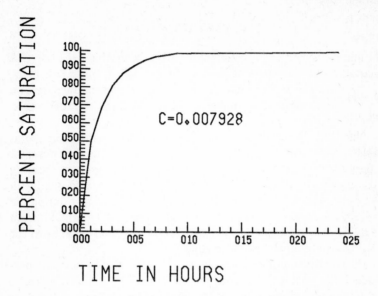

TIME IN HOURS

Fig. 4.3.   A specimen graph produced by the incremental plotter.

on the skin. This potential varies during the course of the contraction of the heart muscle and during the period immediately following the contraction, and the resulting waveform is the electrocardiogram or ECG (Fig. 4). This waveform characteristically contains components characterised by the letters "P" to "T", which correspond to various stages in the cardiac cycle. The study of the amplitudes and shapes of these components and of their relative timing can therefore give some insight into the working of the heart at a particular time. The "P" wave is the clock pulse of the heart, originating as a neuromuscular

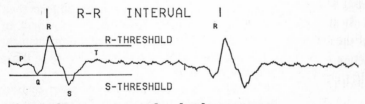

Fig. 4.4.   An electrocardiogram with the components labelled. Two successive complexes, generated by two heartbeats, are shown.

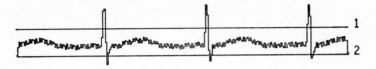

SIMULATED ECG WITH 2 THRESHOLDS

Fig. 4.5. An electrocardiogram with primary definition thresholds set on the CRT display and reproduced on the incremental plotter.

wave of activity radiating from the sino-auricular node near the entrance to the heart from the great veins returning blood to the heart from the body. The large "QRS" complex corresponds to the wave of muscular contraction passing down the central septum of the heart to the apex of the ventricles and up the outer wall of the heart to the exits of the great arteries. The "T" wave indicates the process of recovery of the heart muscle preparatory to another contraction.

Both to the clinician and to the computer, the large "QRS" complex is the obvious landmark for purposes of recognition of the ECG

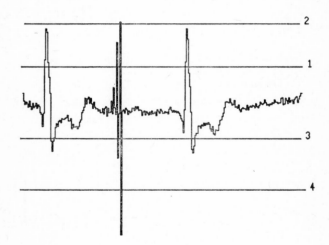

REAL ECG WITH 4 THRESHOLDS

Fig. 4.6. An ECG played back from very thin magnetic tape, with noise from tape imperfections and secondary thresholds set for rejection of noise. Lower-amplitude noise pulses are rejected on the basis of their width.

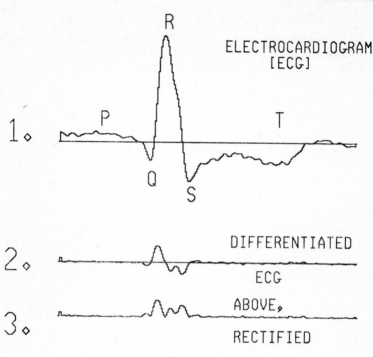

Fig. 4.7.

1. A single, labelled ECG complex with isoelectric line.
2. The same ECG, differentiated.
3. The same ECG, differentiated and then rectified.

This procedure is convenient for measuring the width of the QRS complex.

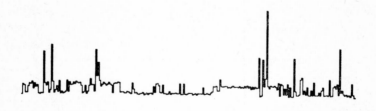

Fig. 4.8. Plot of 256 consecutive R–S intervals measured during a simulated dive.

pattern. The CRT display may be used for setting threshold levels (Fig. 5) for identifying the "R" and "S" waves as changes in the direction of slope after the thresholds have been passed from the right direction. Further thresholds may be set for the rejection of noise artefacts from electrode movement, electrical interference or drop-outs from analogue tape recordings (Fig. 6). This method of pattern recognition is suitable for the measurement of "R–R" intervals (heart rate) and "R–S" intervals, which indicate the rate of conduction of the contractile wave through the ventricular walls.

256 SORTED R-S TIMES.

Fig. 4.9. The same block of R–S intervals as in Fig. 8, but sorted into order of length.

When the "QRS" width is to be studied, however, the derivative of the voltage provides a better landmark for picking up the leading edge of the "QRS" complex (Fig. 7), particularly if it is first rectified, when only a single threshold is necessary for recognition.

The standard tape blocks on the LINC–8 are 256 words long, so it is convenient to collect 256 "R–S" intervals, write them into a block and then collect more. A plot of such a block (Fig. 8) as it stands has a rather ragged appearance, and the distribution of values is not immediately apparent. Furthermore, it presents data from only about four minutes of time. Some of our experiments last for a week, and the prospect of examining thousands of such displays is only slightly less daunting than that of examining a pen recording of a week-long ECG record.

A partial answer to this problem is to sort the "R–S" times into numerical order before display (Fig. 9) or plotting. This produces a sigmoid curve from which the relevant frequencies of long and short times are readily apparent. The main disadvantage is that a given "R–S" time can only be located to within about four minutes of its occurrence, but this is usually unimportant. An advantage is that useful statistics such as the Median, Upper and Lower Quartiles, and so on, always appear in the same position in the display or plot, and are thus very easy to extract.

The overwhelming advantage of sorting before plotting is that it is then possible to plot the data in isometric form without high values at the front obscuring low values at the back (Fig. 10). In this way,

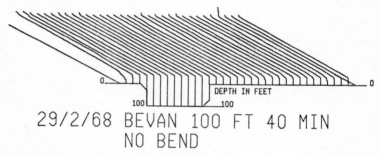

NUMERICALLY-SORTED R-S INTERVALS

DEPTH IN FEET

29/2/68 BEVAN 100 FT 40 MIN NO BEND

Fig. 4.10.   Isometric plot of consecutive blocks of sorted R–S intervals measured during a simulated dive. The front of the plot shows the pressure (in equivalent feet of sea water) at which the first measurement in each block was taken.
Note the absence of significant changes in the distribution of R–S times during the experiment, although the occurrence of two abnormally long R–S intervals may be seen in the fourth block from the end. More than 8,500 individual measurements are displayed in this figure.

many thousands of data points may be plotted in quite a small area in such a way that none are lost and even individual values that are markedly abnormal may be detected. The front of the plot may be tagged with an important experimental variable (such as Pressure in the examples given) so that changes in the distribution of data may be related to the course of the experiment. These graphical procedures have given us a very powerful tool for the detection of slight changes in cardiac physiology over prolonged periods (Fig. 10). Figure 10 presents the data for about four hours, so that this picture tells over 8,500 words in place of the proverbial thousand.

Acknowledgement is given to the Superintendent, RN Physiological Laboratory, for permission to publish this chapter.

Chapter 5

Preliminary Analysis of Blood Flow Characteristics in the Abdominal Aorta by Computer Interpretation

H. Greenfield, Ph.D.
*Assoc. Res. Prof. Computer Science and
Asst. Res. Prof. Surgery
University of Utah*

C. Brauer, B.S.
*Computer Programmer
presently with Information Technology and Systems
Salt Lake City*

K. Reemtsma, M.D.
*Chairman
Department of Surgery
University of Utah Medical Center*

SUMMARY

The relationship of blood flow turbulence to the pathogenesis of atherosclerosis and to the sites of predilection of atherosclerotic lesion formation is being studied by digital computer usage. Fluid dynamic equations are employed in a particular fashion and programmed for optical display. The results are in isometric projection form. The models discussed are: the idealized plaque, the renal section of the abdominal aorta, the abdominal aortic bifurcation, and a combination of the latter two models. All computer formed optical patterns are for the two dimensional steady flow, rigid wall situation;

however, the programming is flexible and future inclusion of pulsatile flow and viscoelastic wall parameters is contemplated. Plots formed from the computer evaluations for plaque positioning and blood flow variances caused by specific vascular geometries are in qualitative agreement with surgical findings. An attempt was made to duplicate the lower abdominal aortic region of a typical young male by computer-optical display experiment and the results are discussed.

INTRODUCTION

A number of investigators [1-3] have suggested a relation of hemo-dynamic factors to the production and distribution of atherosclerotic lesions in the larger vascular channels of the body. Also, it has been noted that atherosclerotic lesions in man occur more frequently in certain regions where branch points, curvatures and bifurcations appear as part of the vascular channel geometry. In turn, as was suggested by Wesolowski et al.[4], a relationship also appears to exist between turbulence and the pathogenesis of atherosclerosis, for, from observations at operations and by arteriography, it has been noticed that turbulence occurs at each site of predilection of atherosclerosis. It was with these thoughts in mind, plus the consideration that the laws of fluid dynamics are applicable to natural conditions in the circulatory system (blood as a fluid is amenable to mathematical approach), that the present investigation was initiated. It is a sub-section of the continuing efforts of the Division of Artificial Organs at the University of Utah to develop such organs for human use. Within the area of study pertaining to the artificial heart the senior author has become involved with the design of an artificial heart valve that allows optimized and correct blood flow. The scope of interest was broadened when a digital computer was requisitioned to develop methods for studying many of such fluid flow problems which pre-viously had been deemed too complicated for analysis. The present investigation is such a problem and fluid flow equations were formed to study blood flow and resulting hemodynamic phenomena in vascu-lar channels of various geometries such as within the renal region and within the area of the abdominal aortic bifurcation. These equations, with boundary conditions specified by physical requirements, were then converted for purposes of solution by the computer and subse-quently processed for optical display purposes and isometric plotting.

MATHEMATICAL MODEL

In pursuing the relationship between turbulence and positions of atherosclerotic plaques (abnormal surface patches within the artery), it should first be briefly mentioned that the motion of fluid is either a streamline one or a turbulent one. In the streamline or laminar case, particles of flow move in definite paths usually parallel to neighboring streamlines. In turbulent motion of a fluid there is an eddying motion which, at any given point, varies from instant to instant. If one places an obstacle within the flow's boundaries, experiments have shown that the fluid behaves in a series of fairly regular changes when flowing about the fixed object in the stream. At low velocity, laminar flow exists about the object up to a point of critical speed where the onslaught of turbulence is noticed. Vortices are shed alternately from the rear top and bottom sections of the immersed object, resulting in an oscillating wake. Further increase in flow velocity causes the vortices to deteriorate into random eddies, signifying fully developed turbulence. Thus, if such a pattern can be duplicated by use of the computer and plotted or displayed optically, a calculated experiment is formed that not only is novel but also allows hemodynamic studies, previously intractable.

The methods employed were of the type initiated by Fromm[5] and Harlow[6], which showed the feasibility of a general numerical solution, and the improved numerical technique as presented by Pearson[7] and by Esch[8]. The numerical technique pursued in the present study was an attempt to solve the general boundary problem. The numerical routines proposed by Pearson was extended to the general finite difference equations and a variable mesh capability was developed. Unlike the Pearson–Fromm programs, the shape of the boundaries within the problem became a variable input. This action facilitated many experiments without changing the main computer program. Finite difference equations for a curved boundary were of a classical derivation as detailed by Forsythe[9], Salvadori[10] and Todd[11]. Application to the present project has been discussed by Greenfield[12]. The treatment of the curved boundary is being reapproached in order to improve its effectiveness. Currently, verification is being attempted.

For the computer experiment then, a mathematical formulation was performed. The obstacle was moved from midstream, where it appears in the general case, and was made a particular case by attaching it to one wall and designing it semihemispherical; an idealized plaque. For other situations, a corner of a side channel entrance, where it attaches to the main artery, was studied. Also, a crotch-form was analyzed as a duplicate of the aortic bifurcation. Both of these shapes were termed obstacles to the blood flow since they caused abrupt changes in the direction of the blood flow and distorted the flow lines. Employing theoretical concepts, resulting fluid dynamic equations were then simplified by the employment of the technique involving required difference equations. Interactive computer algorithms allowed further simplified mathematical forms which were then available for conversion to an intricate computer program. In turn, the finite difference approximations and appropriate boundary conditions formed a solution in terms of a stream function and vorticity function at each point of computational mesh. Subsequent time dependent values at such nodes were then formed by the process of over-relaxation (see Greenfield[13]).

COMPUTER CAPABILITY AND ANCILLARY EQUIPMENT

The computer system used in this study was a Univac 1108 digital computer with a 65K core memory. The secondary storage includes five 'fast' drum units, each serving the program by storing large arrays of data that could not be efficiently held in core memory. Each drum has a temporary storage capacity of 131,000 thirty six bit words. Permanent program and data files are stored on a large 'Fastrand' mass storage drum.

The size of the finite difference computing grid was held at 26×61 cells. This dimensional total was dictated by the physical storage limitations, rather than by computational considerations.

Also included in the facilities was the use of a PDP–8 digital computer that used the Univac 1108 unit's central processor for complicated calculations. The PDP–8 computer was the central control for an IDI optical display unit which was contained in the graphics laboratory. A Sylvania data tablet, the analog device by which coordinate information was obtained by phase measurement, was in the circuit. The user entered alphanumeric data, curve traces, modification data or any form of hand generated graphic data, into the computer through a multiplexing switch to the computer and display scope. The display control multiplexed the stylus position information with computer generated information so that the optical display showed a composite of the pen position output and the computer commanded output. The graphic solutions seen in this chapter, however, were not photographed from the optical display, but rather, were produced by a Gerber mechanical plotter off line, e.g. produced by another computer program after the solution was fashioned on the optical display. The advantages were the obtaining of a hard copy and greater detail.

THE PROGRAMMING APPROACH

The computer program allowed several useful parametric studies as endproducts, each consisting of two plots. The first result contoured the solution space surface while the second was its isometric projection. The plots were:

(a) Stream function—a visual display of the paths of the streamlines in the blood flow. This display can be verified by a laboratory flow visualization technique such as the injection of a dye or radioactive particles into the fluid. The isometric view shows the relative magnitude of the streamlines.

(b) Velocity plots, of which three types were presented, the velocity along the main stream axis, the velocity across the channel, and the combination of both of these as a composite total velocity. The solution space surface plots for each of the velocity components can be obtained experimentally by traversing the space with a hot wire anemometer probe. The isometric view would be tedious to obtain.

(c) Velocity vector plot, which indicated both the direction and relative magnitude of the blood flow. This is analogous to the rather crude laboratory method of visualizing the flow direction by the use of 'streamers' in the flow stream.

(d) The vorticity function plot which indicated the magnitude of the angular momentum of a fluid particle in space. The vorticity function cannot be obtained in the laboratory.

The flow chart indicates the procedure followed (See p. 40).

The subroutine ISOPLT developed for the isometric projection of the solution space is of prime interest since it formed the bottom half

D

of the plots shown in this chapter. This routine is given in the appendix. The complete computer program is too lengthy to be presented. However, to illustrate the insertion of the subroutine ISOPLT into the main program the following assumed equation is used:

$$\sin\left(8*(X-1)/X_L + \tfrac{1}{4}(Y-1)\right) + 1.0$$

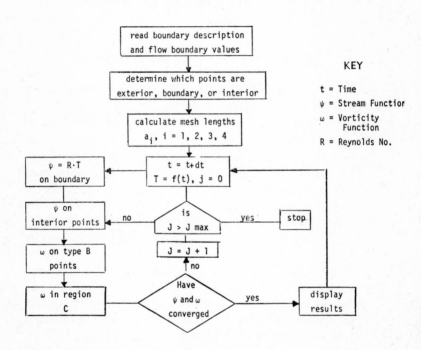

A sample of the calling sequence for the assumed equation is shown in the MAIN program below, as is a listing of the subroutine ISOPLT. This is followed by the surface plot of the assumed equation.

The algorithm[14] used in ISOPLT can be stated as:

(1) Draw the right most line first, moving from right to left.

(2) Lift the pen (blank the display vector) when a line segment drops below (vertical reference) any previously drawn line segment. Similarly, drop the pen, or turn on the display segment when the line segment again moves above the vertical reference line.

Program Variables

TA(I,J) = functional value at the point I,J
M = number of columns in TA
N = number of rows in TA
XP = size of plot, in inches, in horizontal direction
YP = size of plot in vertical direction
BND(I,J) = mode point type

= 0 interior point
= 1 boundary point
= 2 exterior point

A(I,J) = spacing between nodes at (I,J)
A(I + 1,J) = distance from A(I,J) to A(I + 1,J)
.
.
etc.

In the program for which ISOPLT was developed, both A(I,J) and BND(I,J) were calculated from the intersection of a physical boundary description and a mesh grid. In general, however, the arrays A(I,J) and BND(I,J) may be initialized as shown in the sample MAIN program.

@ I FOR MAIN

```
DIMENSION TA(61,26), BND(61,26), A(61,26)    1  BND(I,J)=0
INTEGER BND                                      DO 2 I=1, M
COMMON/D1/CALCMP, TYPE                           BND(I,1)=1
DATA CALCMP/'CALCMP'/                         2  BND(1,N)=1
DATA TYPE/'CALCMP'/                              DO 3 J=1, N
M=61                                             BND(1,J)=1
N=26                                          3  BND(M,J)=1
CALL IDPLOT                                      XP=5.0
CALL PLOT (3.0, 6.0, −3)                         YP=2.5
DO 1 I=1, M                                      CALL ISOPLT
                                                 (M,N,XP,YP,
                                                 TA,BND,A)
DO 1 J=1, N                                      CALL PLOT
                                                 (9.0, −3.0, −3)
X=8.0*FLOAT(I − 1)/FLOAT(M − 1)                  CALL FINI
TA(I,J)=SIN(X+0.25*FLOAT(J−1))+1                 CALL EXIT
A(I,J)=1.0                                       END
```

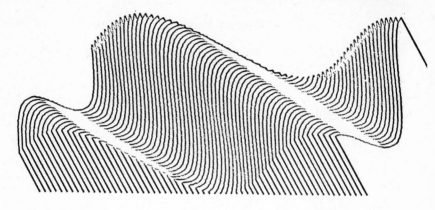

Fig. 5.1. Computer produced solution space surface for
$\sin(8*(X-1)/X_{L_-}^* + \frac{1}{4}(Y-1)) + 1.0$

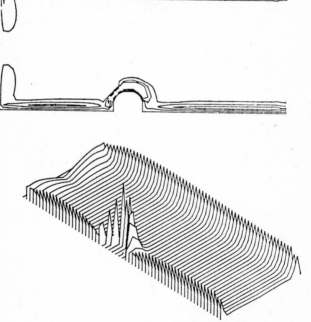

Fig. 5.2.   Idealized atherosclerotic plaque.
VORTICITY FUNCTION
R = 150
Frame No. 10

DISCUSSION

From the results formed by the computer program, for the sake of brevity, only the vorticity function has been emphasized in the sample plots. Also, only one time frame has been chosen from the flow time sequence available. A single blood flow velocity was incorporated into the similarity parameter, the Reynolds number. It is important to note that present results are for conditions of steady blood flow and rigid arterial walls, for the true pulsatile blood motion and the viscoelastic arterial properties are only now presently being investigated. Experimental data, of concern to turbulence as a factor in atherosclerosis pathogenesis, is scarce. However, this phase is also currently being pursued.

Figure 2 shows the vorticity function for an idealized atherosclerotic plaque, attached to the intima of an idealized vascular channel. It is of interest that Bruns et al.[15] discussed the possibility of turbulence inducing high frequency vibrations, which may, in turn, induce a destructive effect upon the arterial wall. Subsequent healing at the turbulent position is viewed as a lesion. If one sees the representation of figure 2 as an interim step in the enlargement of the plaque, the isometric view shows the sharp peaks of turbulence factors and at what points possible vibratory motions should increase as a build-up phenomenon. As Wesolowski et al.[16] mentioned, the second mechanism by which turbulence within the arterial tree might be of importance to the pathogenesis of the lesion is by production of localized and increased lateral wall pressure. Such local hypertension may form a critical injury or result in local accumulation of lipids.

Figure 3, the evaluation of the velocity component along the Y axis of a two dimensional plot, shows increased lateral values for the incremental time frame chosen. Studies of the stages of lipid deposit accumulation, necrosis, and calcification would be aided by continuums of such computer experiments.

Figure 4 shows the stream function study result for the semi-idealized vascular channel in the renal region while Fig. 5 demonstrates the theoretical positioning of vortices (forerunners of full turbulence), formed in the abdominal aortic bifurcation section. Figure 4 shows a marked vortex formed on the distal wall of the upper branching artery; a position where plaques are not uncommon. (Other computer results show heavy vortex formations at other positions when the branching artery is angled differently.) The positioning of heavy vortices at the branch point and immediate sidewalls in Fig. 5 is in direct accord with sites of predilection of atherosclerosic letions as seen by surgical teams. Again, a full continuum motion as presented by time plots is available. In future pulsatile flow studies, at the bifurcation, it is surmised that turbulence that develops at the

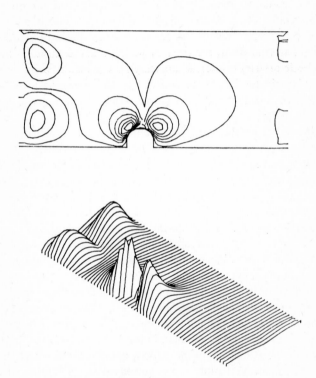

Fig. 5.3.   Idealized atherosclerotic plaque.
V   Velocity
R = 150
Frame No. 10

branch point may be sucked back from the side branches into the main branch during periods of flow deceleration, possibly positioning such turbulence factors further upstream.

The effect of pulsation upon stability of laminar flow is of considerable interest. There is little doubt that oscillating flow will become unstable at Reynolds numbers that are less than the usual critical

Reynolds number (the function of velocity), where turbulence is seen, when these are calculated from the maximum average flow velocities. Evans[17] demonstrated that rapid alterations in the rate of flow will readily cause a break up of the pattern when he employed dye injection as a tracer.

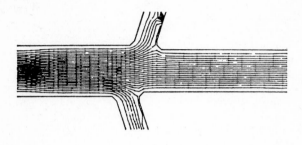

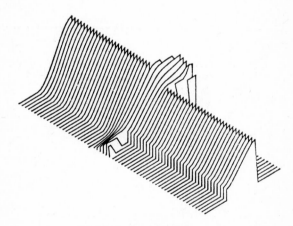

Fig. 5.4.   Renal region.
STREAM FUNCTION
R = 150
Frame No. 15

An attempt was made to duplicate, via the computer-optical display method, the tracing of an aortagram of a young, healthy male, as shown in Fig. 6, for the region extending from the kidney region through to the groin. The obvious assumption might be that the result would be the equivalent of adding Figs. 4 and 5. However, difficulties

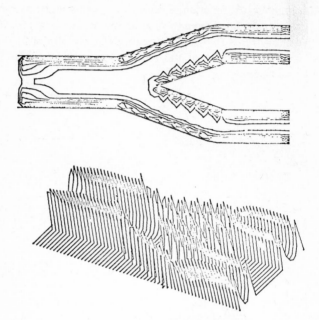

Fig. 5.5.   Abdominal aortic bifurcation.
VORTICITY FUNCTION
R = 150
Frame No. 15

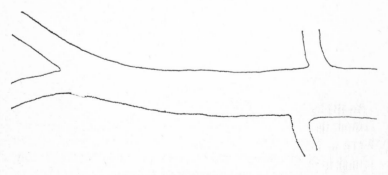

Fig. 5.6.   Tracing of aortogram.
TRACING OF AORTOGRAM of J. R. W.
No. 21–49–70

became evident when the original computer algorithms were utilized. Figure 7, therefore, is not the exact overlay equivalent to Fig. 6.

Briefly, the difficulties encountered are of concern to that portion of the numerical analysis technique wherein a computing mesh is required that covers the section of the arterial tree being studied. This

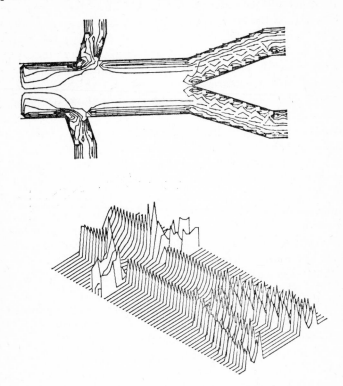

Fig. 5.7.   Combined renal–abdominal aortic bifurcation sections.
VORTICITY FUNCTION
R = 150
Frame No. 15

mesh, or grid overlay, allows required mathematical parameters to be evaluated at numerous points of the flow model in an iterative manner. If there are not enough cells made available in the many-celled grid, the truncation error becomes too large; the truncation error being the error inherent in the use of the integration formulae involved. This error accumulates since the step-by-step procedure within the required numerical procedure compounds the original error, for the procedure is approximate in its initiation. The central aortic section model was thus widened so as to allow more cells and weaken the truncation error. The problem is compounded since there is a maximum amount of grid cells available to the computer program. Too, each renal artery and bifurcation branch of Fig. 7 constitutes an obstacle. At the present, sufficient detail cannot be allocated to each of such areas of flow where detailed study is most required because of computer memory storage capacity. Both of these problems are currently being studied as well as the important task of inserting flexible boundaries. Notwithstanding, the basic approach appears to be sound and the computer program allows discussion of certain qualitative results.

The channel configuration shown in Fig. 7 was also optically displayed in a preliminary form not shown, where the renal side arteries were at complete right angles to the abdominal aorta. No true difference in vorticity configurations could be seen at the bifurcation point when the geometry of the renal area was so changed. The velocity component along the flow axis did show greater deviation at the bifurcation site when the renal side artery was in an angled position, However, this deviation was nullified when both the lateral and axial velocity components were totalled vectorially. Not surprisingly, more turbulence was seen in the renal arteries when the side arteries were angled.

In comparing Fig. 7 with Fig. 5, the vortices in the aortic bifurcation region appeared at greater distances downstream on all walls of the branches in Fig. 7. If one assumes that this shift and the added lengths of vortex motion are due to the part played by the renal section, this may be so. However, it should be remembered that the intermediate distance between the renal and bifurcation areas was foreshortened to fulfil the previously discussed program requirements. It does appear that the vortices' magnitudes were lessened in Fig. 7.

The companion to the vorticity function plot of Fig. 7, the plot of the velocity component along the flow axis (not shown here), showed a change from the typical parabolic front of laminar flow to the blunt turbulent flow front, just past the renal position. It is known that for laminar flow to exist in the lower abdominal aorta a constant flow velocity must be maintained past the side vessels. This constant velocity depends upon a constant ratio of flow to aortic cross-sectional area, however, such a constant ratio does not exist when the side vessel 'bleeds off' part of the flow. The remaining fraction of the flow downstream of a side vessel attempts to fill the downstream main lumen width as if the upstream total flow was still present. The result is a dynamic expansion effect with the usual occurrence of turbulence as seen. As Wesolowski[16] pointed out, the correction for such turbulence is to taper the main channel distal to the side vessel. This will be included as a parameter in future computer programming.

Turbulence, as computed for the side arteries of Fig. 7, commonly occurs when such flow from a main to a side vessel is deflected, allowing distortion of the equipotential lines of flow. In turn, as Texon[18] noted, it can be assumed that in a curved flow resulting from the presence of a side vessel, the net effect of flow forces is an increased pressure on the outer wall curvature and net suction on the inner curvature; the suction effect causing an atherosclerotic response. In the future attempts to incorporate pulsatile blood flow and wall elasticity into the computer experiment, modifications of the response pattern will certainly be seen. However, the basic premise should still exist. As Kravetz[19] pointed out, based upon his experiments utilizing dye injection into glass tubing as side artery models, the effect of the branch vessel was to lower the critical Reynolds number where laminar to turbulent flow transition occurred, by twenty-nine per cent.

Such comparisons of surgical findings and laboratory approaches to the present computer experiments, as discussed in the sections above, merit the continuation of this new approach to hemodynamic studies.

APPENDIX: Subroutine ISOPLT

@ 1 FOR ISOPLT

```
        SUBROUTINE ISOPLT (M, N, XP, YP, TA, BND, A)
        DIMENSION BND (61, 26), A(61, 26)
        DIMENSION TA(61, 26), U(75), V(75), T(75), S(75),
          TEM(75)
        DIMENSION UA(75), VA(75)
        REAL M1, M2
        LOGICAL DOWN, FIRST
        DIMENSIONAL FL(3000)
        COMMON/PLT/SCALE, FL
        COMMON/D1/CALCMP, TYPE
        INTEGER BND, FL, TYPE, CALCMP
```

```
            N1 = N + 1
            N2 = N + 2
            DY = .80*YP/FLOAT(N − 1)
            DX = DY
            X = XP + DX
            ANG = 0.86602540
            ER = 0.2*ANG*DX
            ZSCALE = 0.0
            DO 1 J = 1,N
            DO 1 I = 1,M
            IF (BND(I,J).EQ.2) GO TO 1
            ZSCALE = AMAXI (ZSCALE, ABS(TA(I.J)))
    1       CONTINUE

            DO 10 II = 1,M
            I = M + 1 − II

            X = X − DX
            Y = − DY
            U(1) = ANG*X
            V(1) = −.5*X
            UA(1) = 0.0
            VA(1) = 0.0
            JJ = 1

            NN = N − 1
            DO 2 J = 1,NN
            JJ = JJ + 1
            UA(JJ) = 0.0
            VA(JJ) = 0.0
            Y = Y + DY
            U(JJ) = ANG*(X + Y)
            IF (BND(I,J).EQ.0) GO TO 31
            IF (BND(I,J).EQ.1) GO TO 30
            V(JJ) = .5*(Y − X)
            IF (J.EQ.N) GO TO 2
            IF (BND(I,J + 1).EQ.2) GO TO 2
            UA(JJ) = ANG*DY*(1. − A(I,J + 1))
            VA(JJ) = .5*DY*(1. − A(I,J + 1))
            JJ = JJ + 1
```

```
        U(JJ) = U(JJ − 1)
        V(JJ) = .5*(Y − X) + TA(I,J + 1)/ZSCALE
        UA(JJ) = UA(JJ − 1)
        VA(JJ) = VA(JJ − 1)
        GO TO 2
   31   V(JJ) = .5*(Y − X) + TA(I,J + 1)/ZSCALE
        UA(JJ) = ANG*DY*(1. − A(I,J + 1))
        VA(JJ) = .5*DY*(1. − A(I,J + 1))
        IF (J.EQ.NN) GO TO 22
        IF (BND(I,J + 1).EQ.0) GO TO 2
        IF (J + 2.GT.N) GO TO 2
        IF (BND(I,J + 2).LT.2) GO TO 2

   22   JJ = JJ + 1
        U(JJ) = U(JJ − 1)
        V(JJ) = .5*(Y − X)
        UA(JJ) = UA(JJ − 1)
        VA(JJ) = VA(JJ − 1)
        GO TO 2

   30   IF (BND(I,J + 1).EQ.2) TA(I,J + 1) = 0.0
        V(JJ) = .5*(Y − X) + TA(I,J + 1)/ZSCALE
        UA(JJ) = ANG*DY*(1. − A(I,J + 1))
        VA(JJ) = .5*DY*(1. − A(I,J + 1))
        IF (J.LT.NN) GO TO 2
        JJ = JJ + 1
        U(JJ) = U(JJ − 1)
        V(JJ) = .5*(Y − X)

    2   CONTINUE
        DOWN = .TRUE.

        IF (II.GT.1) GO TO 19
    C   PLOT FIRST LINE WITHOUT HIDDEN LINE
            ALGORITHM. . .
        CALL PLOT1(U(1), V(1), 3)
        DO 21 J = 2,JJ
   21   CALL PLOT1(U(J) − UA(J), V(J) − VA(J), 2)
        DO 23 K = 2,JJ
   23   T(K) = − 10.0
        GO TO 33
```

```
 19   CALL PLOT1 (U(1), V(1), 3)
      DO 20 K = 2,3

 20   CALL PLOT1 (U(K) − UA(K), V(K) − VA(K), 2)
      S(1) = U(1)
      T(1) = V(1)
      FIRST = .FALSE.
      IF (V(4).LT.T(3)) FIRST = .TRUE.

      DO 8 K = 4,JJ
      DO 43 KK = 1,N1
      IF (ABS(S(KK) − U(K)).GT.ER) GO TO 43
      IF (V(K) − T(KK)) 50, 12, 12
 50   IF (DOWN) GO TO 14
      CALL PLOT1 (U(K), V(K), 3)
      DOWN = .FALSE.
      GO TO 8
 43   CONTINUE
      PRINT 44, K
 44   FORMAT (1HO, 30X, 8HHELP...., 2HL = 13)
      RETURN

 12   FIRST = .FALSE.
 13   IF (.NOT.DOWN) GO TO 14
      CALL PLOT1 (U(K) − UA(K), V(K) − VA(K), 2)
      GO TO 8
 14   M1 = (T(KK) − T(KK − 1))/(S(KK) − S(KK − 1))
      IF (ABS(U(K) − U(K − 1)).GT.ER) GO TO 4
      SS = U(K)
      GO TO 5
  4   M2 = (V(K) − V(K − 1))/(U(K) − U(K − 1))
      SS = (M2*U(K − 1) − M1*S(KK − 1 + T(KK − 1) −
           V(K − 1) )/(M2 − M1)
  5   TT = M1*(SS − S(KK − 1)) + T(KK − 1)
      IF (S(KK − 1) − ER.GT.SS .OR. SS.GT.S(KK) + ER)
           GO TO 16
      IF (DOWN) GO TO 27
      CALL PLOT1 (SS, TT, 3)
      CALL PLOT1 (U(K) − UA(K), V(K) − VA(K), 2)
```

```
      DOWN = .TRUE.
      GO TO 8
27    CALL PLOT1 (SS − UA(K), TT, 2)
16    CALL PLOT1 (U(K) − UA(K), V(K) − VA(K), 3)
      DOWN = .FALSE.

8     CONTINUE

33    S(2) = U(1)
      DO 40 K = 2,N
      DO 45 J = K,JJ
      IF (U(J).GT.S(K) + ER) GO TO 40
45    CONTINUE
40    S(K + 1) = U(J)
C         STORE THE MAXIMUM VALUE OF THE V(K)
              LINE IN ARRAY T(K)
      T(1) = V(1)
      DO 41 K = 2,N1
      TEM(K) = − 10.0
      DO 42 KK = K,JJ
      IF (U(KK) − S(K)) 42, 51, 42
51    TEM(K) = AMAX1 (V(KK), T(K − 1), TEM(K))
42    CONTINUE
41    CONTINUE
      DO 11 K = 2,N1
11    T(K) = TEM(K)
6     CONTINUE
10    CONTINUE

      IF (TYPE.NE.CALCMP) CALL SNDFLE(FL)

      RETURN

      SUBROUTINE PLOT1 (A,B,KK)
      DIMENSION FL(3000)
      COMMON/D1/CALCMP, TYPE
      COMMON/PLT/SCALE, FL
      COMMON/PLT2/ISW,IX1,IY1
      INTEGER TYPE, CALCMP,FL
```

```
      IF (TYPE.NE.CALCMP) GO TO 1

      CALL PLOT (A,B,KK)
      GO TO 6

  1   IF (KK.LT.0) GO TO 6
      IX = 70.0*A + 300.0
      IY = 70.0*B = 600.0
      GO TO (2,2,3), KK
  2   IF (KK.EQ.ISW) GO TO 7
      CALL LNTYPE(0)
      CALL LINE(FL,IX1,IY1,IA)
      CALL LNTYPE(3)
      ISW = 2
  7   CALL LINE(FL,IX,IY,IA)
      GO TO 6
  3   ISW = 3
      IX1 = IX
      IY1 = IY

  6   RETURN
      END
```

This research was sponsored by the Public Health Service under NIH grant HE 12202 01A1. Also sponsored by the Advanced Research Projects Agency, Dept. of Defense; monitored by AFSC, Research and Technology Div., Rome Air Dev. Center, Griffiss AFB. N.Y. under contract AF30(602)-4277.

REFERENCES

1. Gutstein, W. H., Lataillade, J. N. and Lewis, L. *Role of Vasoconstriction in Experimental Arteriosclerosis.* Circulation Research, vol. 10, 1962, p. 925.
2. Scharfstein, H., Gutstein, W. H. and Lewis, L. *Changes of Boundary Layer Flow in Model Systems.* Circulation Research, vol. 13, 1968, p. 580.
3. Fox, J. A. and Hugh, A. E. *Localization of Atheroma: A Theory Based on Boundary Layer Separation.* British Heart Journal, vol. 28, May 1966, p. 388.

4. Wesolowski, S. A., Sauvage, L. R., Pinc, R. S. and Fries, C. C. *Dynamics of Flow in Graft Disproportions and in Normal Blood Vessels.* Surgical Forum, vol. 6, 1956, p. 223.

5. Fromm, J. F. *A Method for Computing Nonsteady Incompressible Viscous Fluid Flows.* Los Alamos Scientific Lab. A.E.C. Report LA–2910, 1963.

6. Harlow, F. H. and Welsh, J. E. *Numerical Calculation of Time-Dependent Viscous Incompressible Flow.* Physics of Fluids, vol. 8, 1965, p. 2182.

7. Pearson, C. E. *A Computational Method for Viscous Flow Problems.* Journal of Fluid Mechanics, vol. 21, 1965, p. 611.

8. Esch, R. E. *An Alternative Method of Handling Boundary Conditions and Various Computational Experiments in the Numerical Solution of Viscous Flow Problems.* Sperry Rand Res. Ctr., Sudbury, Mass. SRRC–RR–64–64, 1964.

9. Forsythe, G. E. and Wasow, W. R. *Finite Difference Methods for Partial Differential Equations.* J. Wiley & Sons, Inc., N.Y., 1960.

10. Salvadori, M. G. and Baron, M. L. *Numerical Methods in Engineering* Prentice-Hall, Inc., Englewood Cliffs, N.J., 1961.

11. Todd, J. (ed.). *Survey of Numerical Analysis.* McGraw-Hill, Inc., N.Y., 1962.

12. Greenfield, H. and Reemtsma, *Atherosclerotic Lesion Formation Studies by Computer Experimentation.* Circulation Research, to be published.

13. Greenfield, H. and Brauer, C. *Hemodynamic Studies Involving a Computer Simulation Technique,* presented at International Symposium of Computer Aided Design, Southampton, England, April 1969 (Inst. of Electrical Engrs. Publication No. 51).

14. Brauer, C. *A Computer Program to Plot an Isometric Projection of a Solution Space Surface.* Univ. of Utah Computer Science Dept. internal report, TR–4–9, Aug. 1968.

15. Bruns, D. L., Connolly, J. E., Holman, E. and Stafer, R. G. *Experimental Observations on Post-Stenotic Dilatation.* J. Thorac. Cardiov. Surgery, vol. 38, 1959, p. 662.

16. Wesolowski, S. A., Fries, C. C., Sabine, A. M. and Sawyer, P. N. *Turbulence, Internal Injury and Atherosclerosis.* Biophysical Mechanisms in Vascular Homeostasis and Intravascular Thrombosis, ed. P. N. Sawyer, Appleton-Century-Crofts, N.Y., 1965.

17. Evans, R. L. *On the Mechanism of Turbulent Flow in Liquid.* Proc. 4th Midwestern Conf. on Fluid Mech. Res. Bull. Purdue Exp. Sta. vol. 28 1955, p. 235.

18. Texon, M. *Mechanical Factors Involved in Atherosclerosis, Atherosclerotic Vascular Disease,* eds. Brest, A. N. and Moyer, J. H., Appleton-Century-Crofts, N.Y., 1966.

19. Kravetz, L. J. *The Effect of Vessel Branching on Haemodynamic Stability.* Physics of Med. and Biol., vol. 10, No. 3, 1965, p. 417.

E

Chapter 6

Computer Graphics in Molecular Biology

Lou Katz and Cyrus Levinthal
Graffidi Lab.
Graphics Facility for Interactive Displays
Department of Biological Sciences
Columbia University
Morningside Heights, New York 10027 U.S.A.

ABSTRACT

We are applying computer display techniques to research work on the three-dimensional structure of macromolecules, biologically interesting smaller molecules, and to the three-dimensional neural anatomy of small invertebrates. In order to understand the precise nature of the interactions between molecules and the relationships between structural elements of living organisms, one needs techniques for presenting a description of three-dimensional objects in a convenient form. Since the parameters of interest are often ill defined and sometimes unknown beforehand, visualization of a three-dimensional abstraction of the physical object or system seems to be the most useful technique available to us at present. Construction of three-dimensional physical models has been the traditional method in this approach, although stereoscopic diagrams have also been used. Physical models are severely limited when the molecule or structure becomes highly complex. In that case only a few models are constructed and often only a useful representation of the final stage of the analysis is actually made, while intermediate models are simply not done.

We have constructed a computerized system which makes it psychologically and physically easier to construct pictures for representing three-dimensional structures. The pictures themselves are drawn of successive line segments, and pictures of over five thousand line segments plus alphanumeric characters can be observed as a flicker-free presentation. These images may be rotated, translated and scaled in real time. The real-time user manipulation provides the necessary three-dimensional illusion for perception of complex three-dimensional shapes.

A system for constructing a 'stick' model of any arbitrary molecule based on standard geometrical values for the atoms is described below. This presentation of what is essentially a tree-structure conveys information about the chemical inter-relatedness of the several atoms in the structure and often represents the *a priori* knowledge of the investigator. Methods by which the initial conformation of a molecule possessing internal degrees of flexibility can be changed to provide chemically plausible alternative configurations are also discussed. Since the physical extent of the atoms requires that they do not 'bump' or occupy the same space at the same time, the user can interactively manipulate his model so as to remove contact violations if they occur.

Construction of surfaces of constant electron density, analogous to standard contour maps of X-ray crystallographers is also described. This is a representation of the experimental information, and portrays the space-occupying aspects of the model. It is especially powerful when chemical groups fail to lie in crystallographically convenient directions. Both the stick model and the density surfaces can have text material added (labels). These labels can be attached to specific lines in the picture, and will rotate along with the image. The interactive system for fitting any proposed stick model of a molecule to the observed positions in three-dimensional space of the density maxima enables the structure solution to be verified and conveys what the molecule 'looks like'.

The construction and use of an interactive program which enables the user to record and redisplay the three-dimensional geometry and connectivity of any tree-like structure is described. Its application to the recording and mapping of the neural anatomy of a microscopic invertebrate is discussed.

INTRODUCTION

The precise description of structure on a microscopic and on a molecular level represents a major concern of the contemporary biologist. The aim of this description is to reach an understanding of biological function and mechanism, for it is clear that the three-dimensional state of a molecule provides clues and sometimes explanations for its activity. A protein molecule contains several thousand atoms, and an adequate description of systems of this sort has been our goal. We need to perceive not only the global form of our molecule, but also to discern relationships between its parts, and between it and its environment which may be extremely difficult to formulate precisely. We have constructed a system for representing the data visually, leaving the complex pattern recognition to the observer. This chapter is primarily concerned with describing the computerized system and a few of its applications.

In order to represent a complex three-dimensional structure adequately, a system is needed which can provide the visual clues necessary for three-dimensional perception. The response time of the system must be short, as the orientation of the object on the screen will have to be varied by the user in order to present useful views. In general a very satisfactory illusion of three-dimensionality can be achieved if the object rotates in real-time. The display of the image, with rotation, should be flicker-free, otherwise fatigue will greatly limit its usefulness. These images represent a quantative abstraction of a molecule or a biological system which can contain thousands of atoms or sub-units. Since some of the parameters of the system can be subjected to mathematical manipulation, there is also a requirement for computational power and for two-way interaction between image and digital computer. The implementation of these requirements will be described below.

HARDWARE

The system we have constructed at Columbia University was shaped by the above criteria. For the display function, we have an Adage, Inc. AGT/50 Graphics Display Computer. The AGT is connected to the Columbia University Computer Center's IBM 360/91 via a parallel-wire High-Speed LINK to provide interaction with the complex computational functions.

The AGT itself provides the special hardware: the digital-to-analog converters and an analog matrix multiplier (the hybrid array) to process the output from a vector generator, which provides the real-time rotation of the vectors in the image as it is displayed in a 12 inch CRT. The vector-drawing strategy itself is a natural mode of representation of 'stick' figures. To control the display, there are programatically testable 'variable control dials' which can give continuously variable parameters for rotation, translation or scaling of the images or subimages. There is also a bank of buttons ('function switches') which can be tested for controlling program flow. Efficient production of alphanumerics is provided by a hardware character generator.

Since these displays are to be viewed and manipulated in real-time they require hardware availability measured in hours rather than in seconds. This function would impose an undue burden on the central computer of any organization, so that the AGT itself was selected to provide a stand-alone capability for image display. The numerical computations on the experimental data such as least-squares refinement of X-ray data, or on theoretically developed parameters such as calculation of conformations with minimum energy, require the computational power of the University's IBM/360, and are done there. In order to be able to process display data and carry on display functions without the necessity for intervention by the IBM/360, we provided for local storage on magnetic discs. The size of our images, which often contain more than 5,000 vectors dictates a minimum core size of 16,000 words. A teletype operator console, one magnetic tape drive and a high speed paper-tape reader complete the AGT system. This display hardware itself runs on an interrupt-service basis which allows control and computation programs to run in the computer at the same time as the image is being displayed.

Communication between the AGT and the IBM/360 is over a parallel wire LINK of about 1,000 feet. The 32 bit 360 word is mapped into the 30 bit AGT word by suppressing two bits. Transmission is on a demand basis, and is of the order of 30,000 words per second. The transmission hardware was designed and built at Columbia University. At the 360 end of the LINK, access is through a parallel-data adapter on a 2701 unit attached to a multiplex or channel. In operations under MVT, a small control program in a partition in the 360 handles the communication.

In the description of the system that follows, the division of computation functions between the AGT and the IBM/360 often represents decisions of convenience, rather than a fundamental design philosophy, and many of the functions could have been done (and may yet be done) on the other machine.

Hard copy output from the AGT is, of course, in the form of images. Two systems are presently available. First is an auxiliary oscilloscope with permanently mounted polaroid camera for fast, medium resolution still pictures. Second is photography from the main screen, which gives finer photographs, and through a computer controlled motion picture camera, a record of the three-dimensional images in rotation. Stereo views are also available.

THE DATA STRUCTURE

The description of many diverse three-dimensional models can be accomplished in a common data structure. This eases the process of inter-computer communications and can simplify the process of construction and manipulation of models. The data structure enables us to convey the three-dimensional structure of the model conveniently by creating pictures of all, or of any selected subset of it. We also may wish to adjust parameters of the model for some specialized purpose, such as to minimize its energy and we should be able to compare two or more models to assess their similarities or differences. These tasks are also facilitated by the data structure.

Each object is composed of a collection of units connected together in a well-defined way. Specification of this linkage (connectivity) is the description of the secondary structure. The units can be composed of subunits, etc., down to the most primative units, 'the atoms', whose specification is the primary structure.

Once constructed, a given unit can be stored in a form independent of the method of generation and itself becomes available as a subunit in a higher level construction. The three-dimensional arrangement of the units constitutes the tertiary structure whose description is the goal of this work.

In the case of proteins, the units are the amino acids which occur as twenty different types each with a different side group but containing the same peptide backbone. The peptides form a linear polymer whose connectivity defines the primary structure of the protein. Input to this scheme is a character string which specifies the units in order, and output is a completed data structure. We create our pictures with lines connecting primitive units (atoms) which are numbered. The data structure itself refers to the atoms. This is the expression of the fact that our system contains a primary structure determining feature (the covalent chemical bonds) whose connectivity is invariant.

The lines connecting the atoms are specified by listing the co-ordinates (x, y, z) of each atom and also the number of the FROM atom (the atom from which the bond comes). This is sufficient to describe a TREE structure. For molecules we will treat the lengths of these bonds as invariant, but this is not a fundamental restriction. For loops or rings we introduce the concept of the 'virtual atom' which is an entry in the atom list corresponding to the atom which closes a ring. The first atom of a structure has its FROM listed as 0. Disjoint structures such as two separate molecules can be treated using this notation.

To alter the model we must be able to refer to and change internal parameters. Each atom has a pointer to TYPE, a list of its properties, which is not in general derivable. We also allow for the inclusion of three-dimensional structures such as methyl groups in the atom list.

For molecules we will allow rotation about the bonds joining the atoms. Each atom will thus have an angle associated with it which is called THETA. ANGLO is the nearest atom down the connectivity list whose rotation angle can alter the position of the atom in question. It is obvious that the distance between two atoms is constant if they have the same ANGLO. It is convenient to be able to change which angles are allowed to vary and which are to be treated as fixed so that rigid substructures can be designated. This of course may affect the values of ANGLO for the atoms. Since the number of atom–atom interactions is $\dfrac{N(N - 1)}{2}$ for an atom ensemble, this information is used to reduce the number of interaction terms to be calculated. The interaction energy between two atoms whose separation cannot be changed is a constant which need not be considered when calculating the derivative of the energy.

For those properties which depend on secondary structure a pointer (AANUM) is maintained giving the substructure number (amino acid number). Since restrictions, simplifications or inter-actions may also be imposed which cannot be derived from the con-nectivity, a variable QBACK is also associated with each atom. This gives the first (or last) atom backwards on the list for which a property is to be calculated or ignored.

To summarize, the following set of data is used for describing each ATOM (thing). This set was not designed to be compact or to be efficient in its use of computer storage space. For each atom we will store the following:

1. X coordinate in three-dimensional space.

2. Y coordinate in three-dimensional space.

3. Z coordinate in three-dimensional space.

4. FROM—the atom from which this atom comes via a covalent bond. This array describes the connectivity of the structure.

5. LEVEL—this describes how far out a point is on the branched tree structure.

6. TYPE—atom type. Atom may be points, i.e. real atomic nuclei (carbon, nitrogen, hydrogen, etc.) or a three-dimensional object like methyl groups.

7. VIRT—a special pointer for virtual atoms. For virtual atoms used in ring closure this will point to the real atom it corresponds to. If the virtual atom represents an unsatisfied bond it is a 'vanishing virtual'. These atoms are removed from the structure when two units are joined by connecting respective unsatisfied bonds. Various types of 'vanishing virtuals' are maintained to correspond to chemical valence (single bond, double bond, etc.).

8. THETA—this is the rotation angle about the bond leading from the FROM atom to the present atom.

9. LOCK—a non-zero value for this parameter means that the bond cannot be rotated either for chemical reasons or because the user wishes to maintain the integrity of an already defined structure.

10. ANGLO—the next lowest rotatable THETA which can vary the position of this atom. If two atoms have the same ANGLO then the distance between them is constant.

11. AANUM—the number of amino acid or other subunit that this atom came from.

12. REF—this parameter is non-zero if the atom is a central atom of an amino acid or other predefined substructure, i.e. the atom which connects the substructure to the backbone. This can serve as a reference point for determining various display and calculation options.

13. QBACK—the number of the first atom backwards on the list for which the interaction energy should be included when calculating the energy by using atom pairs. This array is used in general for defining properties of the interaction which cannot be derived from connectivity alone and which may differ depending on the physics of the problem itself.

CONSTRUCTION OF MOLECULAR IMAGES— CHEMGRAFF

A 'stick' model of a molecule represents the connectivity information about the atoms in three-dimensional space. This representation is composed of lines which connect the centers of atoms which are chemically bonded to each other. Several examples are shown in the film. The representation is reasonably transparent—one can view the inside of a molecule, as it is not greatly obscured by atoms on the outside. A crude view of the space occupied by the molecule is presented, but it is very difficult to see on a detailed level whether atoms are reasonably close from this model alone.

Construction of this model on the 360 uses the CHEMGRAFF programs and starts with the character string input of its subunits. For example, the strong "CH3. CH2. OH" will result in the construction of ethanol. In the case of proteins, the subunits themselves, the amino acids, have been previously constructed and stored so that all that is required is the amino acid sequence, such as "VAL. LEU. ARG. . . .". This string is scanned and the subunits are selected and the connectivity arrays completed by the program MOLE.

There are several ways to specify the geometry of the finished molecule. If there is no *a prioro* information at all, the coordinates in the predefined templates are used and standard rotations about the FROM bonds are taken. Each subunits is transformed in order, from its own local frame of reference to the global frame of the molecule by translating it to its point of attachment and rotating it so that its incoming bond and the outgoing bond are of the previously constructed fragment of the molecule are co-linear.

If rotation angles about the bonds are known, they can be applied at this stage. Alternatively, if actual spatial coordinates of atoms are given, they can be substituted.

When the connectivity is completed, the tree must be renumbered to its proper form. Each subtree is uprooted and renumbered by BUNYAN. This removes the distinction between preformed units and finished molecules, and allows the finished molecules to be stored and used as templates in subsequent constructions. At this point all algorithms which perform connectivity-dependent functions will operate correctly and the molecule is ready for display or interactive manipulation.

Display lists are generated by WIRE from the data structure, with options as to detail being implemented with reference to the properties TYPE, REF, LEVEL and AANUM. Among the options one can select to display every atom, or just the atoms of selected amino acids, or a simplified backbone by drawing only from the reference atoms of one amino acid to the next. Labelling information is supplied by SHINGL, and is also keyed for several levels of detail and selectivity.

The output of WIRE is transmitted to the AGT over the LINK (or via magnetic tape if direct interaction is not needed), and the resultant image displayed directly. If desired, the image may also be saved on the AGT disk for later viewing.

MANIPULATION OF THE MOLECULAR MODEL

The algorithm for altering the conformation of the molecules is implemented in a routine, JIGGL, which has two versions. On the 360, the transformations necessary to rotate about any chemical bond are performed digitally, and the transformed coordinates replace the old coordinates in the data structure. In the AGT, the display list is slightly restructured, and a new transformation matrix is calculated for each display frame, using the current values of the rotation angles as sampled from variable control dials. In either case, the proper ordering of the tree structure defines the range of the JIGGL to end when one moves to an atom whose FROM is equal to or less than the FROM of the starting atom. After JIGGLing on the AGT, new values for angles are returned to the 360, and an updated structure is constructed with atom bumps removed.

ANALYTICAL PROCEDURES

The conformation of a molecule is determined by both inter- and intra-molecular forces. The routine SOLVE calls various ENERGY functions to calculate the potential energy of the molecule, and to minimize this energy by altering the molecule conformation. The motions allowed in the present version are only rotations about single bonds—the rotatableness of a bond being indicated in the attribute LOCK. The conceptually simplest energy function is the van der Waals energy. This contains a weak attractive potential when atoms are moderately close and a strong repulsion when the atoms 'bump'. SOLVE calculates the current value of the energy and the derivatives of the energy with respect to the variable angles. Using these derivatives, it takes a small step in the direction of lower energy and then recycles. This routine will go to a local minimum from any starting configuration, but may not find a global minimum. It will assure that atom bumps are eliminated, thus giving a stereo-chemically reasonable model.

DISPLAY SOFTWARE

The interface between the image list in the core of the AGT and the hardware is a software package supplied by Adage, Inc., the 'Display Operator'. This program establishes the interrupt structure, initializes the interrupt pivots and starts the display hardware. It also calculates some of the transformation matrices and loads them into the hybrid array as required.

The PICTOGRAFF programs interpret image information from various sources (LINK, packed 7-track tape, unpacked 7-track tape), format the image into a proper display list and turn on the display. They implement scaling, rotation and stereo viewing of the entire image, and contain labelling and title options. To advance to the next image, a function switch is pressed. Synchrony between the display and the computer-controlled motion picture camera is effected here. Also available is an automatic sequential mode of operation, which automatically advances to the next image after a short time. This mode is particularly useful for viewing a sequence of images representing successive stages of some process.

SPACE FILLING REPRESENTATIONS— ISOGRAFF

Atoms are not points in space. They occupy an extended volume, and many molecular properties are immediately influenced by this fact. Considering a molecule to be composed of hard, impenetrable sphere atoms is a necessary first approximation in understanding the molecule's function. To see if two molecules can interact, it is necessary to show that the reactive atoms can get close to each other and that other atoms do not get in the way.

One space filling form of representation has been created by drawing a sphere about each atom's center. The size of the sphere is an adjustable parameter, so that several options are available. Small spheres reduce the psychological importance of the stick figures and emphasize the location of the atoms, without destroying the connectivity information. At the other end of the scale, spheres drawn to the full van der Waals size (the hard-sphere size) completely obscure the connectivity, but define the available and excluded volumes. The insides of molecules in this representation are hidden. Intermediate sizes permit parts of the connectivity to be perceived, and also sweep out enough volume to convey shape.

The experimental data of the X-ray crystallographer provide the most accurate and detailed information as to molecular configuration. These data result in a description of the density of matter in the crystal (the 'electron density'). Even in the most favorable circumstances, these density 'maps' contain spurious peaks of density, and ambiguous regions. Part of the task of the crystallographer is to identify peaks of density with atoms so that the connectivity of the molecule is preserved and chemical distances and angles are satisfied. The perceptual problem for the user is in determining possible positions for atoms, a task performed well from a picture, but with great difficulty digitally.

Presentation of the X-ray data is done by the three-dimensional contouring package ISOGRAFF. Surfaces of constant electron density are displayed in three-dimensions, at levels which can be selected by the user. Contouring at high values gives a representation of 'mountain tops'—and may show the location of atoms, while lower levels makes connectivity clearer. Usually one level at a time suffices, but occasionally a two-level display is used.

The contouring procedure is performed in several steps. First, the electron density values of each point of a discrete three-dimensional grid is calculated from the primary X-ray data. This calculation is performed on the 360 and results are written out on magnetic tape or sent over the LINK as values on a set of sheets. Up to 128 sheets of up to 4,096 points per sheet can be accommodated by subsequent steps.

In the SETUP step, these values are packed three to a word and stored on the AGT disk. The grid of points represents evenly spaced intervals along the crystallographic cell axes, which need not be orthogonal. In the next step, the parameters of the crystal cell are entered and the coordinate transformation from crystal space to the orthogonal system of the AGT is calculated by MPICK. Since one may not wish to display the entire crystal cell (this is always the case for large structure such as proteins), the limits of the subgrid to be contoured are also entered. Then the PICK is performed. The original data are read, unpacked and the subgrid of points selected. For later efficiency in contouring, three new arrays, a set of x–y sheets, a set of y–z sheets, and a set of z–x sheets are created.

Contouring of the subcell is done by the DRIVE program. Based on input parameters, it creates a display list for contours at a given level, in any (or all) of the three sheet orientations. Linear interpolation along cell edges is used to find the location where a desired contour passes into or out of the smallest grid cell, and straight-line segments are drawn between these points. All the contour curves in one sheet form a single list for later display convenience. This process can be repeated as often as desired, producing maps at different levels. The display lists are stored on disks.

The RETRV program is used to display the results. Previously created lists may be read in and displayed a plane at a time, or all at once. Several lists may be superimposed to show several levels of density at once. If the crystal has been contoured in planes parallel to each of the major crystal planes, and all three sets of contours displayed simultaneously, we achieve the unique result. The net of curves intersect in space to give the illusion of a surface. The curvature of these surfaces usually allows one to locate their centers in three-dimensions.

Since the images are three-dimensional, molecular features that do not happen to lie on convenient planes are not discriminated against, but in fact show up just as clearly as those that do. This has an important application in locating hydrogen atoms in small structures, when the atom itself may not appear in the same section as the large atom to which it is bonded. A clearer view of the actual shape and volume of the atoms, as they are distorted by chemical forces and atomic vibrations is also afforded.

COMBINING CONNECTIVITY AND SPATIAL INFORMATION

It is here that the interactive system for superimposing the stick figure of the molecule and its electron density map provides the extremely powerful technique for structure solution and verification. Simultaneous display of both images, with the stick figure being adjustable in translation and rotation with respect to the map (scaling is usually precalculated), and with full internal rotation (see JIGGL above) provides the mechanism of fitting the model to the map.

Unlike physical models, strict rigidity of the structure is maintained, and results which are misleading due to inadvertent bending or distortion of the physical model are avoided. It is also here where ease of construction leads one to explore many models and to try alternate solutions. The precise molecular parameters are immediately available for comparison with crystallographic parameters to verify the correctness of the model.

COMPUTER SIMULATION

An interesting application of three-dimensional display techniques was made in conjunction with computer simulation of the unwinding of a helical polymer (DNA). The equations of motion for a linked-bead model were applied for several thousand cycles. The three-dimensional coordinates of the links were written out on to magnetic tape as a display list after each iteration. By playing these images back, and varying the viewing aspect, a conceptualization of the unwinding process was obtained that was not apparent from statistical analysis of the data. The AGT was then used to transfer the images to film, to give a motion picture of the process. Thus a sequential process was transformed into a real-time viewable process where human pattern recognition was able to discern inherent mechanism.

SPACE FILLING AND FITTING

Early stages of crystallographic analysis require a different sort of display capability. Plausable models based on a gross shape can be verified or rejected on the basis of space filling considerations alone. Most crystals contain several units related to each other by symmetry elements. A package of programs has been written to place any fundamental unit in a crystal cell, along with all its symmetry-related counterparts. The units have the rotational and translational freedom appropriate to the crystal. By adjusting these parameters, one can easily see if the several units will fit, or if they bump, or even inter-penetrate. A simple example for the case of $P2_1$ symmetry has been included in the motion picture film.

INTERACTIVE CONSTRUCTION OF CONNECTIVITY GRAPHS—NERVE NET TRACING

The graphics display system can also be used as a three-dimensional notebook. This provides us with an alternative method for introducing the connectivity information into the data structure. Using a hand-held device, the 'mouse', which contains three program sampleable buttons, and two dials at right angles to each other, the user can draw straight line segments on the screen. Pressing one of the buttons denotes a branch point. One can then proceed out of a branch, recording other branches on the way. Upon reaching the end of a branch, a second button is pressed, which records this fact, and the program automatically returns to the last unsatisfied node. Trees of any degree of complexity can be recorded this way.

If coordinate information is also known, this can be recorded at the same time as node information. In the two-dimensional case, a photograph of the net to be digitized is projected on to the display screen of the AGT. A spot is moved on the screen under the control of the mouse. When the spot is superimposed on a feature which is to be digitized the buttons are pressed which result in the storage of coordinate as well as connectivity information. This procedure allows for the extraction of abstract connectivity and geometry from photo-graphs of complex organisms, and the replay of the abstract image free of the 'noise' and distraction of other parts of the structure. The application of this program for mapping the three-dimensional neural anatomy of small organisms has been developed in our laboratory and will be described in detail elsewhere.

SUMMARY

We have described in this chapter an interactive computer graphics system for construction, display and manipulation of three-dimensional images. The principal application of this system in our laboratory is for molecular model building and structure determination. Here, the ease of construction and accuracy of manipulation of the complex three-dimensional forms are the important parameters in its utilization. The feature of the system that makes it work satisfactorily is the ability to display a 'rotatable' image of many thousand line segments, so that the effective three-dimensionality through this user controlled rotation allows perception of form and space necessary for problem solution.

ACKNOWLEDGEMENTS

1 Programmers for Version I of CHEMGRAFF (1965–1968)
   "PACKAG"   Steven A. Ward
                  Andrew T. Pawlikowski
                  Martin Zwick
                  Hal Murray
   "CHEMPK"   C. David Barry
2 Programmers of Version I of ISOGRAFF
                  David Avron
3 Programmers of Version II of CHEMGRAFF (1969–1970)
                  Wendy H. Raskind
                  Andrew T. Pawlikowski
4 Systems Programming and Version II of ISOGRAFF
                  Reidar J. Bornholdt
5 Nerve Mapping and Interactive Connectivity Input
                  Randle Ware
6 Simulation of DNA Unwinding
                  Elliot Simon

REFERENCES

1. Levinthal, C., Barry, C. D., Ward, S. A. and Zwick, M. *Computer Graphics in Macromolecular Chemistry.*
2. Secrest, D. and Nievergelt, J. (eds.). In *Emerging Concepts in Computer Graphics.* W. A. Benjamin Inc. (1968), 231–253.
3. Levinthal, C. *Molecular Model Building by Computer.* Scientific American (1966) **214**, 42–52.
4. Supported by Special Research Resource Grant FR 00442, and Contracts PH 43–67–1131 and PH 43–69–1019 from National Institutes of Health.

Chapter 7

Interactive Computer-Generated Stereoscopic Displays for Biomedical Research

Part I:
Development of Techniques

J. White and J. Perkins
National Institute for Medical Research
The Ridgeway
N.W.7

The N.I.M.R. hybrid computer system consisting of a TR48 analogue computer, DDP 516 digital computer with 8K of store and an interface with analogue/digital and digital/analogue conversion was intended for experimental computing with two main areas of application, simulation and on-line to experiments. For the simulation of biomedical systems, the models first need to be established and at this conceptual stage some interaction by the research worker with the computation is required. Thus immediately meaningful visual displays were needed which were also necessary for the on-line applications. This led to the development of a program for co-ordinate plotting which could provide displays in two dimensions on to a cathode ray tube or $x–y$ recorder.

A difficulty for biomedical workers wishing to use simulation techniques, is the need to provide a mathematical representation of the system. Visual displays opened up the possibility of using available data to provide the nucleus of the display and to be able to insert other features intuitively as a guide to the ultimate system. Applications of this kind could arise in electron microscopy where thin slices are used

71

to build up the solid structure and in molecular biology for determining the atomic structure of a molecule.

By drawing successive two-dimensional projections of a structure as it was rotated about its x, y or z axes, effective three-dimensional displays could be obtained. The orientation could be under program control or adjusted manually to provide an interactive display. With the capability to rotate and translate the display it was possible to provide a stereo pair for direct observation as a 3D structure which could then be rotated for any desired projection. Further refinements were to add perspective and to provide captions for the atoms.

The displays were normally viewed on a cathode ray screen through mirrors but could be transferred to an analogue X–Y recorder to provide a permanent record which could be viewed in the same way.

TECHNIQUES

Three-dimensional structures may be represented on a 2D plane by projecting the x, y and z co-ordinate points on to that plane. To display a different view, all the co-ordinate points need to be multiplied by a transformation matrix and the new computed co-ordinates plotted. The structure could be rotated about the axes by adjusting potentiometers to produce voltages proportional to the angles of rotation. These voltages were passed through an A/D coverter into the digital computer which calculated the transformations and passed the transformed co-ordinates via D/A converters to the display, so providing an interactive display.

By duplicating the display, adjusting separation by another potentiometer and transforming the right-hand display about the y-axis, a stereo pair was produced which could be orientated into any position. Perspective transformations could be similarly applied.

The data was stored in a display file as co-ordinate sextets (x_1, y_1, z_1, x_2, y_2, z_2) defining the end points of the constituent line segments. These sextets were picked up sequentially by the main program, transformed to the resultant x^1, y^1 and x^1, y^1 values which were passed to the vector generation sub-routine for drawing.

All transformations referred to the basic data so the data file remained unaltered to avoid corruption of the data due to summation of the rounding errors, were the data to be continuously transformed.

Part II :
Applications

D. A. Franklin

Computer Unit for Medical Sciences
St. Bartholomew's Hospital
E.C.1

CRYSTALLOGRAPHY

The original impetus behind the adaptation of the original non-stereoscopic display into a fully stereoscopic system was the wish to give crystallographers an aid in the visualisation of molecular structures. These are essentially three-dimensional, and their representation by the usual two-dimensional isometric or planar methods is often unsatisfactory. I might add that we were not pioneering in this. Johnson's ORTEP program[2] produces such images on a graph-plotter, using a FORTRAN IV program with a 32K word machine, though this is a fixed display, and graph-plotted stereo pairs have been produced at Professor Hodgkin's laboratory at Oxford. In fact, stereo pairs are rapidly becoming the standard way of publishing crystallographic and molecular biological results.[1] Our plan to use an interactive CRT display was not new in crystallography either; Levinthal[3] describes the use of a two-dimensional interactive display to investigate chemical reactions and to show how compounds can be assembled from subunits. However, as far as we know, the combination of interactive display with full stereoscopy has not been used before for crystallographic work, especially on a computer of this size.

For those who are not familiar with X-ray crystallography, I will outline briefly the situation. A crystal of the material being investigated is mounted so that it can rotate inside a cylindrical photographic plate which almost envelops it, and a narrow pencil of X-rays is directed at the rotating crystal (Fig. 1). In this situation the crystal behaves as a three-dimensional diffracting grating; the diffracted X-rays hit the photographic plate and produce spots of various intensities on the (developed) plate (Fig. 2). From the intensities of

F*

the spots and their positions on the plate, it is possible to work out
the architecture of the molecule of the crystalline substance and
obtain the relative positions of the atoms of the molecule as three-
dimensional coordinates in a suitable frame of reference. It is from
these coordinates that we can draw a 'stick diagram', as it is called,
of the molecule on the CRT tube or graph plotter. In our case, the
X-ray crystallography is carried out at the Chemical Laboratories of

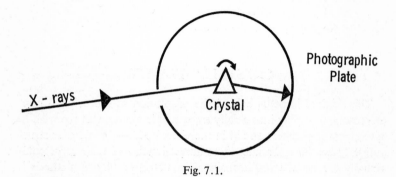

Fig. 7.1.

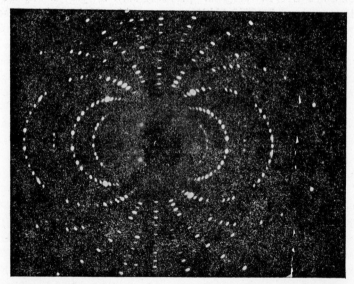

Fig. 7.2. Provided by the Crystallography Group, University of
Cambridge Chemical Laboratories.

the University of Cambridge by Dr. Olga Kennard and Dr. David Watson and their colleagues, and preliminary computer analysis of the results takes place on the IBM 360/44 at the Institute of Theoretical Astronomy, Cambridge. The results of this are passed to Mr. Terry Scott at the Medical Research Council Computer Services Centre, whose 'Unique Molecule' program completes the working out, on the London University Atlas, of the three-dimensional coordinates of the atoms in the molecule and also gives a list of the *connectivities*, i.e. a statement of which atoms are joined to which. This information is the raw data for the stereo system, along with any notes as to which

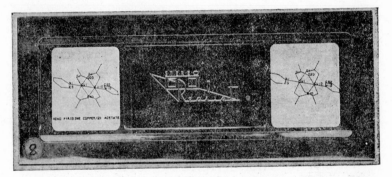

Fig. 7.3. Stereoscopic slide viewer.

atoms are to be labelled, and how. The whole thing has been rather a ribbon development and clearly could be much more integrated, although it would not be possible to carry out the crystallographic analyses on the DDP 516 system we are using.

For all its clumsiness, this prototype system and the one now current (described in Part III of this chapter) have proved of use to the crystallographers at Cambridge and elsewhere. I must emphasise that they are the ones who really use the system, not us. They come along when the data has been punched, they manipulate the controls to provide the view or views which they consider best for their purposes and when they are satisfied the graph plotter is switched in to provide a permanent record. The system is user-oriented, and this has always been the aim.

The drawings can be viewed by a mirror-system as has been described, but a very satisfactory method for mass-viewing is the stereoscopic slide (Fig. 3), with a stereo pair projected through crossed polarising filters for lenses. A screen is also needed with a special type of surface, but the paraphernalia is worth it, I think, because the results can be really stunning.

One interesting by-product of all this is that it has enabled us to find errors in published data. Crystallographic and (bio-) chemical literature very often contains lists of the coordinates of atoms in various molecules and their interconnections, but the difficulties in drawing the structures sometimes seem to lead to unnoticed mistakes in printing; we have found one or two published structures which on being examined by the stereo display system have proved to have errors in them. We suggest that a system of the kind we are describing could help prevent such errors from getting into the literature and would in any case provide quicker and improved visualisation of molecular structures.

Crystallography is the only discipline so far where the system is in active use as a research tool. However, we have looked into one or two other possibilities, and the prospects are promising.

MATHEMATICAL FUNCTIONS

One of the more familiar uses of stereoscopy, and one which tends to feature rather prominently in the advertising material of graph-plotter manufacturers, is the representation of surfaces as functions of two variables. Naturally, we have tried our hand at this as well, though without the refinements of 'hidden-line' deletion and so on— I think this tends to lose information, and in any case, to see the whole structure as though it was made of wire is often very illuminating, especially as we can rotate the figures to get the best view (Fig. 4).

I would suggest that there is a problem to solve here, one which artists have had to face ever since painting began; namely, how do you represent a surface in two dimensions? What guide marks must be drawn so as to lead eye and brain into recognising that a surface is being displayed? Mathematicians tend to indicate by drawing contour lines, as we and others have done, but artists use other cues, for example, highlighting and the use of light/shadow contrasts. I think that computer graphics needs to develop similar techniques, or at least techniques for similar purposes.

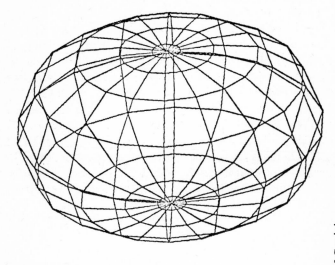

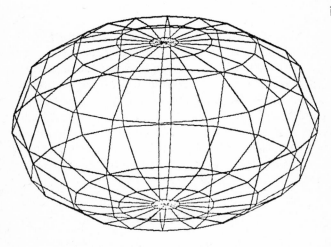

Fig. 7.4. Ovoid of Cassini.

RADIOTHERAPY

One medical application where the stereoscopic presentation of results would seem to offer considerable benefits is in radiotherapy. Computers are already being used to display or print dosage distributions in radiation treatment planning, but this is either two-dimensional—showing the distribution of radiation dose in a plane cross-section through the patient—or as masses of figures for three-dimensional dosage distributions, which the radiotherapist and radiation physicist find very difficult to handle. Isometric projection with rotation has also been used to simulate three-dimensional distributions, but as far as I have been able to find out, stereoscopy has not. As an example of what it can do, consider this picture of an *isodose surface* for an unscreened irradiated platinum iridium wire (Fig. 5).

This is a surface of revolution of the curve

$$\frac{K\theta}{h} = d_p \tag{1}$$

about its longitudinal axis, where d_p is the dose at a point P in rads, θ is the angle subtended at P by the wire, h is the perpendicular distance of the point P from the axis of the wire, and K is a constant incorporating various factors of the wire such as its length and radioactivity. The curve shown is one along which at any point the dose is 0.9 rads, and the surface is therefore one of which the dose at any point is also 0.9 rads.

Stereo has seemed to me for a long time to be a natural vehicle for three-dimensional radiation treatment planning, although it must be admitted that the distribution shown is for a very simple and idealised situation; a real-life distribution from, e.g., 30 needles would be a more difficult undertaking. The formula (1) on which the distributions shown are based was provided by Dr. Gordon Jamieson of the Middlesex Hospital, whose advice and guidance in this application I am very happy to acknowledge, and I am also grateful to Mr. John Clifton, of University College Hospital, for some very helpful discussions.

OTHER POSSIBILITIES

Allowing for the team's natural enthusiasm for this work, I think that its possibilities are unlimited and I will stick my neck out and

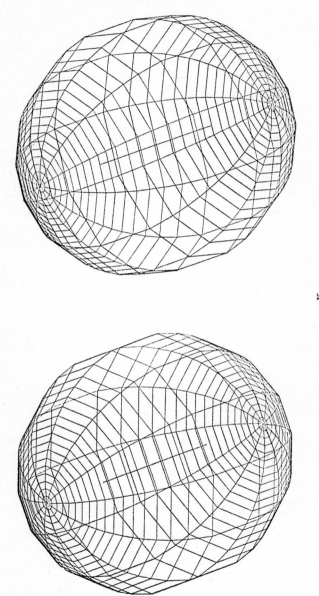

Fig. 7.5. Isodose surface $\frac{K}{h} = d_P (K = 12, d_P = 0.9)$.

say that I am convinced that interactive computer-generated stereo-scopy is one of the most important developments to appear on the computer scene for some time. I am not saying this because we happen to have done it—other people almost certainly have done so as well—but because as time has gone on and we have had a chance to explore this tool, there have been so many facets of its application which have become apparent. For instance, certain visual effects we have noticed while viewing structures have made me suspect that there is an application here in the psychology and physiology of vision. The ability to produce and reproduce precisely defined structures, deform-

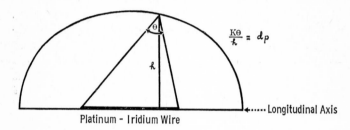

Fig. 7.6.

able in a precisely definable way, must I am sure be of use in this field; I think the whole question of visual cues could be profitably investi-gated using a computer-driven stereo system.

Again, in a different area altogether, why should hospital building proposals not be evaluated this way? Building layouts and architec-ture could be examined before a brick was laid or a decision taken to lay one, with buildings being examined from every viewpoint without the necessity of making models. I believe there is an archi-tectural development of some sort in graphics under way in the United States, but I have not seen any details published.*

SOME POINTS ARISING IN PRACTICE

I would like to conclude by bringing a couple of practical points to your attention. Perhaps they are obvious—in which case I ask readers' forgiveness for raising them—but they were not so to us at the time, and perhaps they might help one or two people.

* Eds. note: see architectural section in "Advanced Computer Graphics", Plenum, 1970.

Firstly, it seems to me that a graph plotter is essential in developing CRT display systems. It is about the only way I can think of for ensuring that your program is working properly. For instance, during the development of the programs it was while watching the graph-plotter drawing at slow speed that we realised that a program error was causing the diagram to be drawn twice over in a single cycle of the program. This kind of error cannot be detected by the CRT visual display alone, because things are happening too quickly, but of course it limits very severely the complexity of the diagrams which can be drawn before regeneration flicker becomes intolerable.

Secondly, I would advise anyone who wants to get involved in stereoscopy to talk to the photographers. Stereoscopy has an honourable history in that profession, and they have been producing stereo-scopic stills and films for nearly 100 years. They know a great deal about it, and in fact it was out of a casual conversation with Richard Barlby of the Photographic Laboratory at the National Institute for Medical Research that the system for generating our images arose. We had been mentally preparing ourselves to getting down to sorting out the mathematics of stereoscopic transformations so as to produce the stereo pairs; Richard happened to remark that photographers often took stereo pairs with 'mono' cameras by simply moving the camera round the object and taking pictures from different positions. This made it obvious that all we had to do on the computer was to project two planar images slightly rotated with respect to one another and there would be our stereo pair. It took John White exactly an afternoon to change his two-dimensional projection program into a three-dimensional stereo system as a result of this remark of Richard's. We found out later that other people had also thought of this— 'there is nothing new under the Sun', it says in Ecclesiastes—but as far as we were concerned, the idea came from photography, and we have continued to benefit from the advice of Mr. Barlby and his chief, Mr. Cyril Sutton.

ACKNOWLEDGEMENTS

This has been an interdisciplinary effort, as is so common with computer applications in medicine or biology. Apart from the people whose help has been already acknowledged, we are very grateful to Dr. Olga Kennard and Dr. David Watson of Cambridge University

Chemical Laboratories, who guided us in the crystallographic work along with Mr. Terry Scott of the Medical Research Council Computer Services Centre. Although I myself am no longer with the MRC, this work was carried out while I was employed by the MRC Computer Services Centre and a visitor to the National Institute for Medical Research and I would like to thank my ex-chief, Mr. B. K. Kelly of Computer Services, and Mr. J. Perkins of the NIMR for giving me the opportunity to work in this fascinating field and for their support and encouragement.

REFERENCES

1. Freeman, H. C. in *Advances in Protein Chemistry*, ed. Anfinsen, C. B., Anson, M. L., Edsall, J. T. and Richards, F. M. Academic Press, 1967.
2. Johnson, C. K. *ORTEP: A Fortran Thermal-Ellipsoid Plot Program for Crystal Structure Illustrations*. Oak Ridge National Laboratory, June 1965.
3. Levinthal, C. in *Emerging concepts in Computer Graphics*, ed. Secrest, E. and Nievergelt, J. W. A. Benjamin Inc., 1968.

Part III :

Development of the System for Molecular Structures

E. Piper, J. Perkins and F. G. Tattam
National Institute for Medical Research
The Ridgeway
N.W.7

One immediate application of the interactive display is to provide the illustrations for a reference work on crystallographic structures being produced by Dr. Olga Kennard, of the Department of Organic Chemistry, University of Cambridge. Over a thousand structures are to be referenced and these need to be orientated to specific positions with appropriate captions.

A stereo pair can be displayed on the cathode ray tube for each molecule, orientated to a suitable position by the crystallographer and then transferred to the Calcomp incremental plotter for a permanent record, suitable for reproduction.

A Fortran program is used to input data to the display routine. This permits the co-ordinates from the 'unique molecule program' output to be used directly, avoiding data preparation and error checking. This program finds the mean and centres the co-ordinates before they are stored so that the molecule can be rotated symmetrically. During this process, the co-ordinates are also scaled, making the largest one tenth of full scale to enable a ten-to-one size change to be obtained.

In the original drawing program for general application, the start and end point of each line to be drawn were supplied. In the case of the molecules, the co-ordinates are normally produced as a table of atom x, y, z co-ordinates accompanied by a list of inter-connections. This way of describing the drawing required has a number of advantages for our purpose. The co-ordinates for each atom are only written down once resulting in less storage and making access of the co-

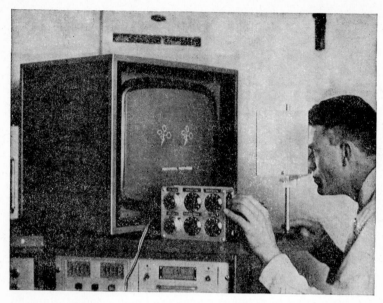

Fig. 7.7.   CRT display and interactive control unit.

Fig. 7.8.   Interactive control unit.

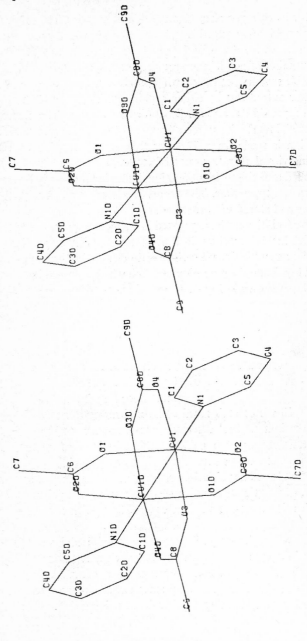

MONO PYRIDINE COPPER (II) ACETATE - PYCUAC01

Fig. 7.9. *X–Y* plot with captions.

ordinates for labelling or alteration considerably easier. The original connectivity list is re-arranged by a 'sort' program which produces a set of instructions for drawing the molecule in such a way that the minimum number of pen lifts are required. This procedure results in a considerable speeding of the plotting process.

The plotter output can be either a pair of 5″ × 5″ or 10″ × 10″ drawings to provide a standardised display format. The format includes fiducial marks for stereo alignment of the final reproduction and provision is made for a 72 letter title and description on the diagram. Individual atoms, groups of atoms or the whole molecule can be labelled to facilitate identification, by typing the required atom numbers on the typewriter. The actual labels are read in with the co-ordinates and stored until required.

The above facilities are incorporated in a single program where the input routine has been written in Fortran and the display routine in the computer assembler language. The cathode ray viewing procedure is shown in Fig. 7 and the potentiometer unit which allows adjustment of the display is shown in Fig. 8. A typical output of the incremental recorder is given in Fig. 9, which is a stereo pair of mono-pyridine copper (II) acetate with complete labelling of atoms.

Index

A

Abdominal aorta 35–55
Adage, Inc. AGT/50 computer display system 58
Advanced Research Projects Agency, Dept. of Defense 54
AFSC, Res. and Tech. Division 54
Airway pressure monitoring 10, 12
Alarm signals 12, 14, 15
Analog displays 9, 10, 12, 15
Arterial pressure 10
Atherosclerosis 35
ATLAS 75

B

Barlby, Richard 81
Beaumont, J. O. 17
Bentley, R. E. 19
Biomedical research 71–86
Blood flow characteristics 35–55
Bolt, Baranek and Newman 20
Branching questionnaire technique 2–8, 12
Brauer, C. 35
Bruno, D. L. 43
Burning, tubes 10, 14

C

CalComp (California Computer Products Inc.) 10, 83
Cambridge University 75, 83
Cardiac catheterization labs 9
Charter sizes 14
Chart recorder 10
CHEMGRAFF program 63–64
Clifton, John 78
Clinical measurement and monitoring 9
Clinical notes 1–8
Colour television 16
Columbia University 56
Computer Instrumentation Ltd. 20

Computer Unit for Medical Sciences 73
Connectivity graphs 69
Content independence 6
Control unit, interactive stereo 84
Cope, D. W. 19
Crystallography 73

D

Data collection and retrieval 1–8, 9–18
Data input 2, 21, 83
Data presentation demands 9, 14–17
Data retrieval 5
Department of Health and Social Security, London 1, 9
Digital Equipment Co. Ltd. (DEC) 19, 20, 27
  DEC LINC-8 computer 27
Division of Artificial Organs, Utah University 36

E

Electrocardiogram 10, 29–33
Esch, R. E. 37
Evans, R. L. 45

F

Fail-safe systems 9, 16
Flashing signals 12, 15
Flexibility to vary format 15
Fluid dynamics 35–36
Forsythe, G. E. 37
FORTRAN 73, 83, 85
Franklin, D. A. 73
Fromm, J. F. 37

G

Gerber plotter 39
Gerbode, F. 17

87

Graffidi Lab.   56
Graph plotting   29–34
Greenfield, H.   35

H

Hard copy   10, 14, 16, 22, 25, 39, 42, 60, 81, 85
Hardware   7, 9–13, 14, 20, 27, 29, 38, 58–60, 71–72
Harlow, F. H.   37
Heart Research Institute, Pacific Medical Center, S.F.   9
Heart valve, artificial   36
Hodgkin, Prof.   73
Hospital layouts   80
Hospital Medical Information System 1–8

I

IBM 360/44 computer   75
IBM 360/91 computer   58
IBM 1801 computer   9
IBM Inc.   9
ICL 1905E computer   7
IDI display unit   38
Index display   28
Information systems   1–8, 9–18
Information Technology and Systems 35
Institute of Cancer Research   19
Institute of Theoretical Astronomy, Cambridge   75
Intensive treatment unit   9
Isodose charts and curves   20, 24, 25, 26, 78, 79
Isoelectric line   32
ISOGRAFF package 65–67
Isometric projection   35, 78
ISOPLT subroutine   39–41, 49–54

J

Jamieson, Dr. Gordon   78
JIGGL routine   64

K

Katz, Lou   56
Kelly, B. K.   82
Kennard, Dr. Olga   75, 82, 83

King's College Hospital   1–8
Krovetz, L. J.   49

L

Levinthal, Cyrus   56, 73
Light pens   15
LINK   58
London University ATLAS   75

M

MAIN program   40–41
Manipulation   64
Mathematical functions   76
Mathematical model   37, 71
Measuring   9–18
Medical Research Council Computer Services Centre   75
Middlesex Hospital   78
Milan, J.   19
Models, mathematical   37, 71
Models, physical   56
Models, stick   57, 74
MOLE program   63
Molecular biology applications   56–70
Monitoring   9–18
'Mouse' controls   69

N

National Institute for Medical Research   71
Neuromuscular wave   31
NIH grant   54
Nurses using computer displays   1–8, 10, 11, 12, 14
Nursing routines display   3

O

ORTEP program   73
Osborn, J. J.   17

P

Patient identification   7
Patient monitoring   9–18
Patient registration   2
Pattern correlations   16
PDP-8 computer   7, 20, 38
Pearson   37

Perkins, J.   71, 82, 83
Photographic copying   16, 60
Photographic Lab., NIMR   81
Physiological lab. applications   27
Physiological transducer amplifiers   9
PICTOGRAFF program   65
Piper, E.   83
Plotters   10, 25, 26, 29, 39, 42, 81, 85
Polaroid cameras   16, 60
Position transducers   20, 21
Printing error detection   76
Priority flashing   12, 15
Programmed Console   20
Public Health Service   54
Pulsatile flow   36

Q

Questionnaire   2

R

Radiotherapy applications   19–26, 78
Raison, John C. A.   9
Reemtsma, K.   35
Repetitive displays   14
Reynolds numbers   43, 44, 45, 49
Ridsdale, B.   1
Royal Marsden Hospital   19
Royal Naval Physiological Laboratory   27

S

St. Bartholomew's Hospital, London   73
Salvadori, M. G.   37
Scott, Terry   75, 82
Simulation   6, 35–55, 68, 71
Sino-auricular node   31
SOLVE routine   64
Stereoscopic displays   71–86
Stereoscopic slide viewer   75, 76
Surface representations   76

Sutton, Cyril   81
Sylvania data tablet   38
System commands   6, 13
System response time   7

T

Tattam, F. G.   83
Tektronix   10, 20
Television, closed circuit   10, 16
Television, colour   16
Texon, M.   49
Three-dimensional display   56, 71–86
Threshold levels   33
Todd, J.   37
Touchwires   15
Treatment display   3
Truncation error   47, 48
Tube life   10
Typewriter terminals   2
TV rasters   15

U

'Unique Molecule' program   75
Univac 1108 computer   38
University College Hospital   78
USPHS grant   17
Utah, University of   35

V

Viscoelastic wall parameters   36
Vision, psychology and physiology   80
Voice communication   16
Vorticity function plots   39, 42, 43–54

W

Washington University   20
Watson, Dr. David   75, 82
Wesolowski, S. A.   43, 49
White, J.   71
Wilton-Davies, C. C.   27

DATE DUE

APR 16 1984			
NOV 6 1984			
MAR 9 1987			
MAY 7 1988			
APR 0 1 1991			